AF334250

American Enterprise:

Nineteenth-Century Patent Models

TO ALL TO WHOM THESE LETTERS PATENT SHALL COME:

Whereas Alexander C. Twining of New Haven, Connecticut has alleged that he has invented a new and useful Improvement in Apparatus for making Ice which he states has not been known or used before his application has made oath that he is a citizen of the United States that he does verily believe that he is the original and first inventor or discoverer of the said Improvement and that the same hath not to the best of his knowledge and belief been previously known or used: has paid into the treasury of the **United States** the sum of Thirty five dollars and presented a petition to the **COMMISSIONER of PATENTS** signifying a desire of obtaining an exclusive property in the said Improvement and praying that a patent may be granted for that purpose **These are Therefore** to grant according to law to the said Alexander C. Twining his heirs administrators or assigns for the term of seventeen years from the fifteenth day of April one thousand eight hundred and sixty two the full and exclusive right and liberty of making constructing using and vending to others to be used the said Improvement a description whereof is given in the words of the said Alexander C. Twining in the schedule hereunto annexed and is made a part of these presents

In Testimony whereof I have caused these Letters to be made Patent and the Seal of the **PATENT OFFICE** has been hereunto affixed **GIVEN** under my hand at the City of Washington this fifteenth day of April in the year of our Lord one thousand eight hundred and sixty two and of the **INDEPENDENCE** of the United States of America the eighty sixth

Caleb B. Smith Secretary of the Interior.

D. P. Holloway Commissioner of Patents

Countersigned and Sealed with the Seal of the Patent Office.

American Enterprise:

Nineteenth-Century Patent Models

Cooper-Hewitt Museum

The Smithsonian Institution's National Museum of Design

LC 83-73525
ISBN 0-910503-42-7
Printed in the United States

Design by Katy Homans
Typesetting by Trufont
  Typographers
Printing by Mercantile Printing

Abbreviations

Lenders of the models illustrated are identified as follows:

Dennis Clark   DC
Decker Manufacturing
  Company   DM
Deborah Friedman   DF
Jason Friedman   JF
Petersen Collection   PC
Frederick M. Shelley   FS
Smithsonian Institution   SI
Alexander Catlin Twining   AT

Unless otherwise noted, all
photographs are by Kim
Nielsen, Smithsonian Institution.

front cover:

Improved Sugar Evaporator
John K. Leedy
Bloomington, Illinois
September 16, 1862
Patent No. 36,462
PC

frontispiece:

**Improvement in Apparatus for
  Making Ice**
Alexander C. Twining
New Haven, Connecticut
April 15, 1862
Patent No. 34,993
Inventor's copy of the patent
  papers
AT, Photograph: Bill Jacobson

back cover, left to right:

Improvement in Clothes-Pins
Samuel Reid and
  Lewis M. Berry
Chicago, Illinois
March 3, 1874
Patent No. 148,088
SI

Improvement in Clothespin
George K. Farrington and
  Bradford S. Potter
Bloomington, Illinois
March 9, 1875
Patent No. 160,661
SI

Improved Clothes-Pin
Jeremiah Greenwood
Fitchburg, Massachusetts
November 15, 1864
Patent No. 45,119
SI

Improvement in Clothes-pins
William H. Mayo
Menasha, Wisconsin
March 23, 1875
Patent No. 161,138
SI

Clothes-Pins
Dexter Pierce
Sunapee, New Hampshire
May 25, 1858
Patent No. 20,364
SI

Improvement in Clothes-Pins
Henry Mellish
Walpole, New Hampshire
September 23, 1873
Patent No. 143,024
SI

Contents

Foreword

As a part of celebrating the centenary of Simpson Thacher & Bartlett, we are pleased to sponsor this important survey of nineteenth-century American patent models at the Cooper-Hewitt Museum. From its beginning in 1884, the firm has been closely associated with persons and companies having the inventiveness so aptly illustrated by these models. The triumph of technology in the last one hundred years shows that we in this country and millions around the world are heirs to the ingenious and arresting spirit which these models embody.

Cyrus R. Vance

Introduction

The nineteenth century was a time of staggering growth in our young country as well as one of remarkable achievement. Within a relatively short span of time, the United States was transformed from a rural society into an industrial giant. Progress was closely linked to patriotic pride and ambition, and optimism and energy ran high.

The unprecedented number of patents issued during this period suggests that the search for self-sufficiency and greater comfort was so zealous that inventing became almost a national pastime. Vigorous efforts were directed toward improving and refining— toward devising processes and products that were more efficient, faster, cheaper, and safer.

In the brief decades during which models were a requirement with all patent applications, ingenious machines, devices, and gadgets were invented—in every area of human endeavor, including agriculture, the domestic arts, communications, manufacturing, and transportation. Most of the inventions patented were never produced; of those that were, relatively few were commercial successes. They were generally short-lived, being rendered obsolete by further improvements. Many models for these inventions have survived, however, and collectively they bear witness to the deep-rooted American belief in progress.

The spirit of philanthropy is particularly American, too. The inventor Peter Cooper, who conceived this museum, and industrialist Andrew Carnegie, in whose former home the Cooper-Hewitt is housed, were among the greatest philanthropists of their day. They would be delighted to know that "their" museum is benefiting from a tradition they helped to establish more than a century ago. In an immensely thoughtful gesture, the firm of Simpson Thacher & Bartlett decided to celebrate its hundredth anniversary in a way that would both support a local institution and give pleasure to the public. We are deeply grateful to the members of the firm for allowing us to share this happy occasion through their generous sponsorship of "American Enterprise: Nineteenth-Century Patent Models."

In addition, we acknowledge our great appreciation to the institutions and individuals who loaned objects to the exhibition, the contributors to this publication, and the others who helped us realize this enjoyable undertaking.

Lisa Taylor, Director
Cooper-Hewitt Museum

Acknowledgments

As Guest Curator, Robert C. Post has shared his extensive knowledge of patent models and the patent process in the preparation of both the exhibition and publication.

Douglas E. Evelyn, Deputy Director of the National Museum of American History, not only marshalled the necessary forces for a very large and complicated loan from that museum to the exhibition, but also contributed to the publication with his essay. Kendall Dood and George Nelson expanded the book's scope with their articles from two different and enlightening points of view.

Cliff Petersen, champion and guardian of the largest private collection of patent models, has been a generous and enthusiastic lender. His able assistants in Garrison, New York, are Patricia McMahon, Beverly Cutten, and Steve Novak. We extend our thanks also to the following lenders to the exhibition:

Dennis Clark
The Cooper Union for the
    Advancement of Science
    and Art
Decker Manufacturing
    Company
Deborah Friedman
Jason Friedman
National Archives
National Portrait Gallery,
    Smithsonian Institution
New Haven Colony
    Historical Society
Frederick M. Shelley
Alexander Catlin Twining

This exhibition would not have been possible without the support and assistance of our colleagues at the National Museum of American History, Smithsonian Institution. The early and continuing interest of Roger Kennedy, Director, assured the success of the project from the outset. Gary Kulik, Chairman of the Department of Social and Cultural History, and Arthur Molella, Chairman of the Department of the History of Science and Technology, were our guides to the vast resources at NMAH. Martha Morris, Registrar, supervised the complex loan arrangements. Many loans come from "1876: A Centennial Exhibition," whose Collections Manager is Melodie Kosmacki.

Our thanks are due to Barbara Suit Janssen who as Special Project Manager undertook the coordination of loans from the following Divisions, whose staff members were helpful throughout:

Archive Center:
    Spencer Crew

Division of Ceramics and Glass:
    Susan H. Myers

Division of Community Life:
    Richard E. Ahlborn
    Ellen R. Hughes
    Carl H. Scheele

Division of Costume:
    Claudia B. Kidwell

Division of Domestic Life:
    Jennifer Oka
    Rodris Roth
    Anne Marie Serio

Division of Electricity and
Modern Physics:
    Bernard S. Finn
    Elliot N. Sivowitch

Division of Extractive Industries:
    Francis Gadson
    G. Terry Sharrer
    Robert G. Walther

Division of Graphic Arts:
    Elizabeth M. Harris
    R. Stanley Nelson, Jr.

Division of Mechanical and
Civil Engineering:
    David H. Shayt
    Robert M. Vogel
    William E. Worthington, Jr.

Division of Mechanisms:
    Carlene E. Stephens

Division of Medical Sciences:
    Michael R. Harris
    Everett A. Jackson
    Ramunas Kondratas

Division of Military History:
    Edward C. Ezell
    Harry Hunter

Division of Photographic History:
    Linda Benbow
    Eugene Ostroff

Division of Physical Sciences:
    Jon B. Eklund
    Ann Seeger
    Deborah J. Warner

Division of Textiles:
    Barbara S. Janssen

Division of Transportation:
    John N. Stine
    L. Susan Tolbert
    William L. Withuhn
    John H. White
    Roger B. White

Most of the photographs in the book are by the Office of Printing and Photographic Services, Smithsonian Institution, whose quick and capable work we appreciate: Richard K. Hofmeister, Mary Ellen McCaffrey, Kim Nielsen.

For the benefit of his advice throughout, our thanks to Silvio Bedini at the Smithsonian.

This project resulted, as always, from the collective and creative efforts of Cooper-Hewitt Museum staff. Catalyst for these was Dorothy Twining Globus, Exhibitions Coordinator, who gave the exhibition and publication their direction and final form. She was assisted in research and coordination by Lucy Fellowes. Cordelia Rose, Registrar, negotiated the extensive loans, assisted by Steven Langehough. Robin Parkinson designed the exhibition. Harold Pfister, Assistant Director, was both friend and advisor at every stage. We are grateful also to the many others who helped in countless ways.

For their roles in the realization of the publication, we thank Stephen Frankel, editor, and Katy Homans, designer.

L.T.

Patent Models: Symbols for an Era

Robert C. Post

When the laws relating to the patent process in the United States were recodified in 1836, they included a stipulation that was then unique and remained so. As part of the matter of disclosure, every applicant was required to "furnish a model of his invention, in all cases which admit of representation by a model, of a convenient size to exhibit advantageously its several parts." The law was not explicit as to the actual information content of the model; the term has a diversity of definitions in any event, and the collection the Patent Office subsequently amassed reflected this diversity. Some are classics of the modelmaker's craft, some decidedly homespun; some are abstract, some perfect miniatures. Some isolate the key "several parts," some put the invention into full context—a component into a whole machine, for example, or the machine into a workshop. Some are static, but the great majority include moving parts.

Some of the surviving models remain pristine while others are very much the worse for wear. Many betray the results of overcrowding, inadvertent mauling, and storage under less than optimum conditions, though one frequently finds evidence of maintenance efforts by clerks whose techniques included the application of a distinctive gold paint to stem corrosion. Affixed to most—with a loop of cotton ribbon traditionally dubbed "red tape" (the government also used it for bundling documents together)—is a punched cardboard tag denoting the patentee, the patent number, the date, and the nature of the patented device, a supporting "document" of inestimable value.

In whatever their state, the patent models still in existence constitute a rich artifactual record of one of the most extraordinary chapters in the history of design and invention. This epoch spans the forty-four years between 1836 and 1880, namely from the middle of Andrew Jackson's presidency to the close of Ulysses S. Grant's. Some seven thousand models that had pre-dated the new law were destroyed by fire and relatively few were replaced. The model requirement was actually repealed in 1870 but the Patent Office, by its own rules, kept requiring them until 1880, and they were often submitted even after that.

From our vantage point in a technological milieu commanded by the chip and the laser, the Shuttle and the ICBM, as we approach the twenty-first century, the world of Jackson and Grant—and of Samuel Morse, Samuel Colt, Cyrus McCormick, Richard Hoe, Elias Howe, Charles Goodyear, John Ericsson, Peter Cooper, and similar "men of progress" who moved in that world—may seem more remote than it should. In fact, it is to that mid-nineteenth-century epoch that we may trace many of our most salient national values, for this was the era of America's "take-off" (the term is Walt Whitman Rostow's) into sustained economic growth.

Men of Progress
1862   Oil on canvas
51⅜ x 76¾ in. (130.5 x 195 cm)
Christian Schussele (1824–1879)
National Portrait Gallery,
   Smithsonian Institution,
Transfer from the National
   Gallery of Art; Gift of
   Andrew W. Mellon, 1942

(left to right)
William Morton, 1819-1868
James Bogardus, 1800-1874
Samuel Colt, 1814-1862
Cyrus McCormick, 1809-1884
Joseph Saxton, 1799-1873
Charles Goodyear, 1800-1860
Peter Cooper, 1791-1883
Jordan Mott, 1799-1866
Joseph Henry, 1797-1878
Eliphalet Nott, 1773-1866
John Ericsson, 1803-1889
Frederick Sickles, 1819-1895
Samuel Morse, 1791-1872
Henry Burden, 1791-1871
Richard Hoe, 1812-1886
Erastus Bigelow, 1814-1879
Isaiah Jennings, 1792-1862
Thomas Blanchard, 1788-1864
Elias Howe, 1819-1867

In 1856 Jordan Lawrence Mott, the New York ironware manufacturer, commissioned Christian Schussele to do a painting portraying selected Men of Progress, *which was to include Mott himself and eighteen of his contemporaries. The men were never actually assembled as they are shown; as was customary for such large group portraits, each sat separately for the painting, which took six years to complete. The* Men of Progress, *though all worthies, are a peculiar convocation mainly representing the fancies of Mott, who positioned himself between Peter Cooper and Joseph Henry in the imaginary scene.*

Some of these men have remained among the stalwarts of our textbook accounts of American invention. Several are represented in this book by

models of inventions they patented, including Colt, Cooper, Morse, Howe, Sickels, Blanchard, and Ericsson. Ericsson had not been on Mott's original list, but when his ironclad ship, the Monitor, vanquished the Merrimac at Hampton Roads, the finished painting had not yet been delivered. Schussele superimposed Ericsson's likeness on the drapery beside the central column, and, as another finishing touch, added a portrait of Benjamin Franklin on the wall to the left. Franklin's presence perhaps enhanced the painting's reception as a conclave of great inventors, yet much more remarkable was the inclusion of Franklin's illustrious successor in electrical science (and first secretary of the Smithsonian Institution), Joseph Henry.

Henry was an important inventor, to be sure—of the first electric motor, among other things—but unlike the others, who were generally partisans of the patent system, Henry held no patents. He regarded patents as antithetical to a spirit of free inquiry, as a corrupting sort of self-indulgence to which no "true man of science" would resort. (Franklin never took out a patent either.) Yet Henry is shown in a pose of apparent fellowship with the likes of Samuel Colt, who was famous for his public posturing, and Samuel Morse, who had patented and commercialized devices actually conceived by Henry, and whom Henry detested.

Several of these men were in fact more significant as entrepreneurs than as inventors. Mott himself, a major manufacturer of anthracite stoves,

was not nearly as important as an inventor of devices "for the evolution and management of heat" as the "philosopher of caloric," Eliphalet Nott, shown next to him. Jennings was neither an inventor of consequence nor a successful entrepreneur. Goodyear, while he made an important discovery, was so inept a promoter that he lived in poverty for most of his life, and in fact had died penniless two years before the painting was completed. Saxton designed scientific instruments for the federal government, and, while he had obtained several patents earlier in his career, was fundamentally of the same mind as Joseph Henry.

Mott was surely entitled to his own concept of progress, and of who had "altered the course of contemporary

civilization," though such inventors were in reality far outnumbered by those whose names mean little to anyone today, even to specialists in the history of invention. The notion that a lone inventor can revolutionize society is naive, as is the general concept of so-called technological breakthroughs. The true history of technology is fraught with false starts, dead ends, well-conceived inventions that failed for want of entrepreneurial skills, and "inventors" whose fame and fortune is largely the product of such skills. Most successful inventions have a pre-history, and many are the result of a fortuitous blend of technology, enterprise, and capital resources.

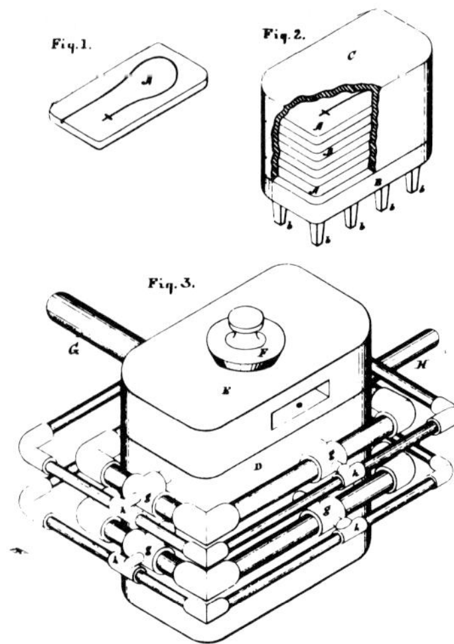

Carbonizer
Thomas A. Edison
Menlo Park, New Jersey
October 18, 1881
Patent No. 248,423
(see also page 73)

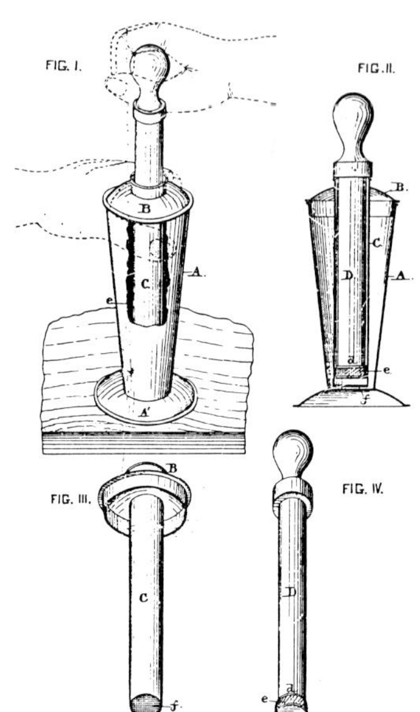

Improvement in Atmospheric and Vacuum Egg-Beaters
Henry E. Marchand
Pittsburgh, Pennsylvania
April 7, 1879
Patent No. 214,936
(see also page 34)

In the mid-1830s the U.S. was moving toward technological self-sufficiency in most realms, leadership in some. The first major industrial exports to Europe date from the 1840s. American manufactures—especially agricultural machinery, machine tools, and precision mechanisms such as clocks, locks, and repeating firearms—first commanded a world stage in 1851 at the Exhibition of the Industry of All Nations, the Crystal Palace in London. Two years later, New Yorkers staged a Crystal Palace exhibition too, with a building that was a frank imitation of the Hyde Park prototype, but with a very different aim: to spotlight "the choicest products of the luxury of the Old World and the most Cunning Devices of the Ingenuity of the New." Americans took liberal advantage of their first international exhibition, and in the realm of "Cunning Devices" a new orchestration unfolded, sufficiently rich to denote the New York Crystal Palace as a turning point in the history of American invention and enterprise.

This change is borne out in the numbers of patents issued. The week that President Franklin Pierce inaugurated the New York Crystal Palace, sixteen patents were granted.

Although in the entire history of the American Republic the cumulative count had not reached 20,000, the total for the preceding year had been 1,014, and since 1836 some 10,000. By the time of the next great world's fair in America—the Centennial Exposition of 1876 in Philadephia—the Patent Office routinely issued more than a thousand patents each month and the cumulative total had multiplied tenfold, topping 200,000. At the time of the World's Columbian Exposition of 1893 in Chicago, more patents were being issued annually than in all the years prior to 1853 combined, and the upward curve would become far steeper still (patent number 1,000,000 was issued in 1911, number 3,000,000 in 1961, and the number is now almost 4.5 million). It was apparent by 1893, or even in 1876, that the U.S. was emerging as the most innovative of nations, the wealthiest, and the premier exporter of manufactured goods—"peacefully working to conquer the world," in the phrase assocated with that epitome of American entrepreneurial prowess, the Singer sewing machine.

As early as the 1840s, the exponential nature of invention in America was obvious to acute observers. In his annual report for 1844, Charles Page, the Patent Office's Chief Examiner, wrote that "Man's wants increase with his progress in knowledge; and hence the paradoxical truth, that the growing number of inventions, instead of filling the measure, increases the capacity." The nation's economic take-off coincided with that increase in capacity and *also* with the era when inventors were required to submit patent models. The latter was a coincidence, beyond doubt, but one which coincidentally endows the models with a relevance to technological and entrepreneurial history every bit as striking as their relevance to the history of design.

The models, it also turns out, collectively constitute a prime symbol of a phenomenon that was long a keynote of American national pride, something called Yankee Ingenuity. More specifically, the models are a testament to an epoch which might be termed America's Mechanical Age. To be sure, there are many species of model that are not primarily or at all mechanical—indeed, here special notice has been paid to domestic furnishings, to ways of embellishing materials, to the functional aspects of heating, cooking, cooling, freezing, gasification, and illumination, and to pastimes— and yet the majority of inventions represented by the models fit somewhere under the rubric of mechanics. In 1876, when Edward H. Knight, editor of the *Official Gazette* of the Patent Office, took several thousand models to Philadelphia to exhibit at the Centennial, he arranged them on the basis of an enchanting taxonomic scheme thought to comprehend the full range of mankind's inventive endeavors:

Agriculture
Architecture
Chemistry
Civil Engineering
Clay
Electricity
Fine Arts
Firearms
Gas
Glass
Harvesters
Heat
Hoisting
Horse Powers
Household
Hydraulics
Ice
Journals and bearings
Leather
Light
Mechanical movements
Metalworking
Metallurgy
Mills and Presses
Navigation
Pneumatics
Printing
Railways
Steam
Stone
Textiles
Vehicles
Woodworking

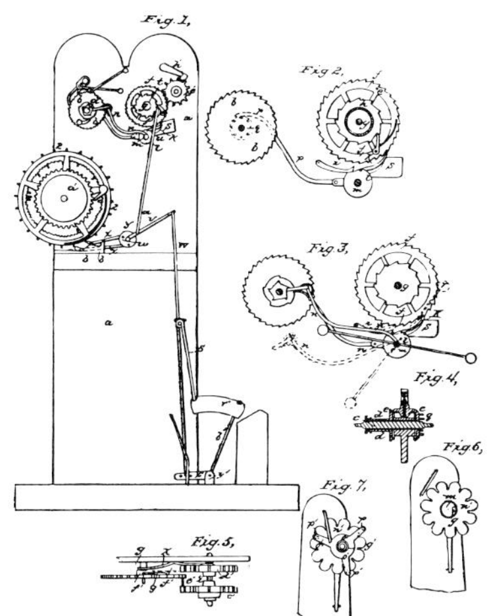

Clock Escapement
Aaron D. Crane
Newark, New Jersey
February 16, 1858
Patent No. 19,351
(see also page 106)

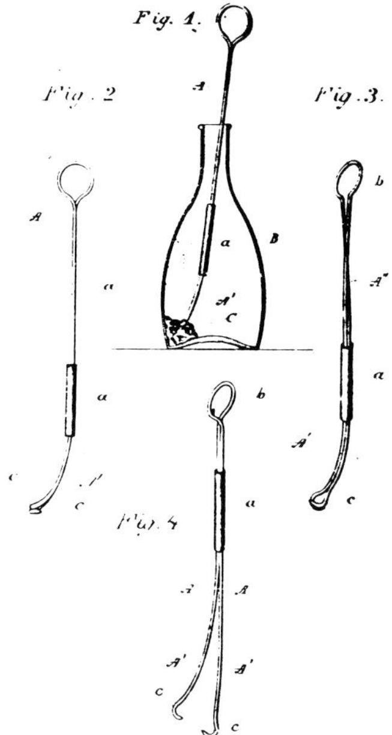

Bottle Washer
Thomas Davies
Poughkeepsie, New York
September 29, 1885
Patent No. 327,071
(see also page 43)

At least half of these categories are explicitly mechanical and all but a handful involve arrangements of conventional materials or components. It is also worth pointing out that the majority of patent applications submitted today would not fit any of these categories.

In the last two decades of the nineteenth century, after the model requirement was rescinded, inventions in the realm of applied electricity, chemical engineering, thermodynamics, and machine control signalled a qualitative shift as significant as the quantitative amplification we have just noted, a shift so important it has been labelled the "Second Industrial Revolution."

If models had not already become superfluous to the process of examining patents, they were now. There has never been a "typical" patent, but if an ordinary one at mid-century was for an improved mower or steam-engine accessory, a churn or a washer, or perhaps a new mode of fitting up an item of furniture, *a fin de siècle* patent was likely to be for something in the realm of circuitry or cybernetics, and to be appropriately represented by a schematic or even a set of equations. Hence patent models, as three-dimensional representations that somehow embody a tangible arrangement of "several parts," are by and large congruent with a discrete era in the history of invention. They are incidentally congruent with the rise of the railroads, the first complex mechanical "system"; when the requirement was terminated the new systems world was the world of electricity, which not only was not mechanical, it was not even visible.

Here it is that one begins to perceive the ultimate significance of the models. As Kendall Dood of the U.S. Patent and Trademark Office has written:

> . . . *the patent model requirement was uniquely suited both to the state of American technology in the mid-nineteenth century and to the popular conception of technology as something entirely and immediately apprehendable through the senses, and it is this lost view of technology which the surviving models most uniquely memorialize.*

This perceptive, even eloquent observation serves also to introduce an important consideration by referring to the *surviving* models. The models are far from a comprehensive record of an inventive epoch, for at least 100,000 have been destroyed (perhaps a great many more than that), while the exact whereabouts of only some 40,000 is known. The rest, conceivably 100,000, are so scattered that a thorough inventory would be virtually impossible. One cannot even be certain of how many there were at key junctures in the past.

The presumption has been that the Commissioner of Patents continued to ask for models only in highly esoteric cases—according to legend they remained mandatory for flying machines and perpetual motion devices—and yet one finds hundreds of models for perfectly ordinary contrivances such as churns and watch escapements bearing tags from the 1880s and 1890s and even past the turn of the century. It can be assumed that custom had a strong hold on some inventors, and that a certain status adhered to having a model on deposit at the Patent Office in Washington. An application was sometimes even accompanied by more than one model and, of far greater consequence, the Patent Office often kept the models from rejected applications. Depending upon the prevailing administrative philosophy regarding "novelty, originality, and utility," between 35 and 65 percent of all petitions for letters patent were turned down. This would suggest the possibility that the office displayed twice as many models as there were patents. It is impossible to know how many models were made but it is quite likely to have been several hundred thousand.

Regardless of how many there are, every one of the patent models is significant, as a model and as a survivor. One definition of model is "a pattern or mode of structure or formation," and a great many patent models stand as perfect embodiments of the techniques by which various sorts of things were fabricated and assembled, and if tagged, are also precisely dated. Many represent artifacts which survive in no other form. All are some kind of art, fine or folk, and all exemplify the extraordinary range of variation in the design process.

Robert C. Post *is editor of* Technology and Culture, *the international quarterly of the Society for the History of Technology, and* Railroad History, *the bulletin of the Railway & Locomotive Historical Society.*

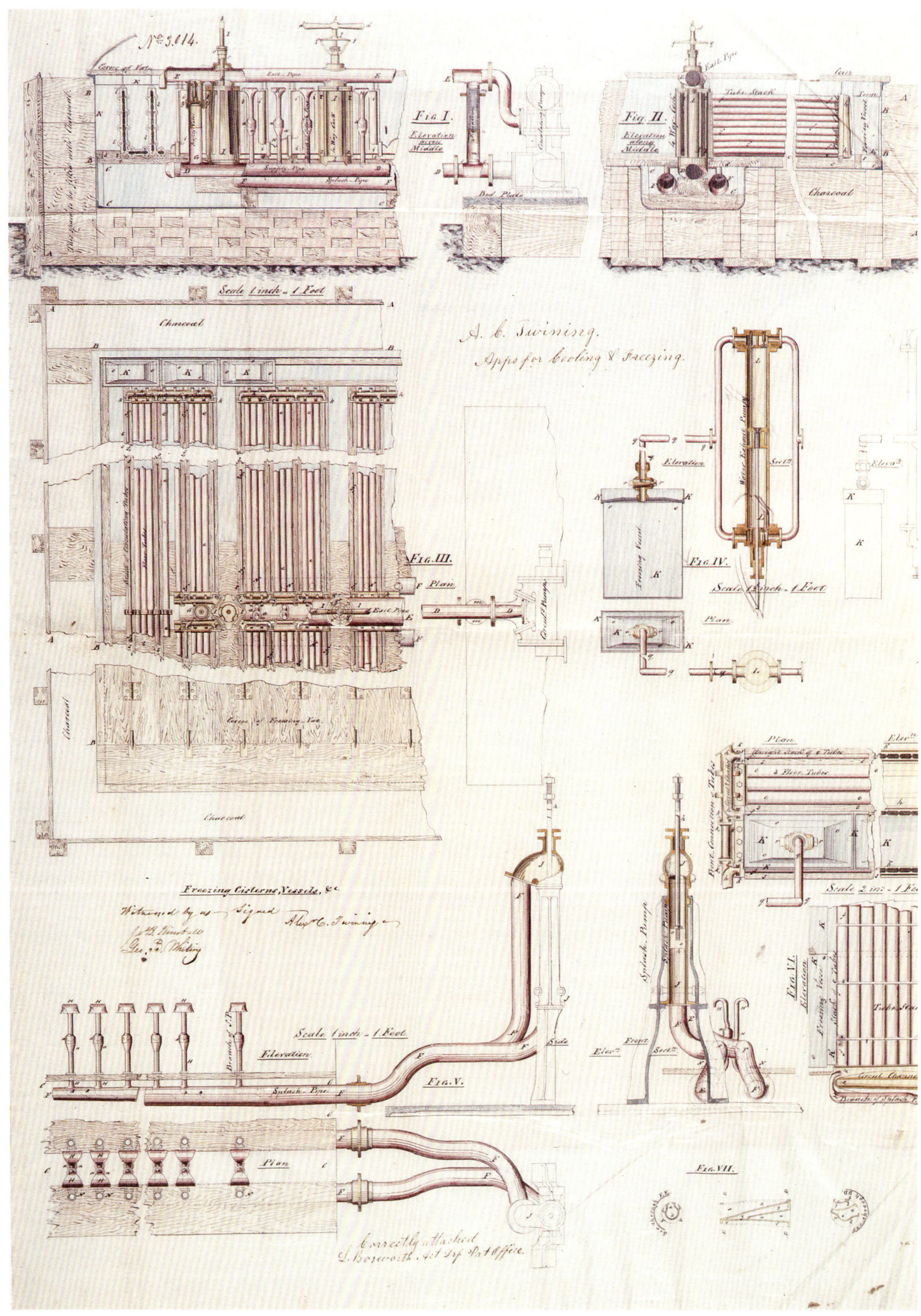

Improvement in Apparatus for Cooling and Freezing
Alexander C. Twining
New Haven, Connecticut
December 20, 1861
Patent No. 34,018
Inventor's copy of the patent
  papers
AT

The Patent Process, 1836–80

To obtain a patent, an inventor had to submit to the Patent Office the following:

1. a specification, a description of the invention rendered in detail sufficient to enable anyone skilled in the requisite arts to duplicate it. This disclosure (now called "enablement") was what the inventor traded in return for receiving a temporary monopoly; otherwise, after the patent had run out the public might not have gained anything in return.

2. a drawing or drawings, complementing the specification and ordinarily keyed to it by means of letters relating specific passages in the disclosure to specific parts of the drawing(s).

3. a model, "not more than twelve inches square . . . neatly made, the name of the inventor should be printed or engraved upon, or affixed to it, in a durable manner." (These technical limitations were not rigidly observed.) There was no disclosure requirement pertaining to the model, nor any formal enumeration of the role it was to play in the process of disclosure.

At first the application fee was $30 for U.S. citizens ($500 for British subjects; $300 for all other aliens—a telling commentary on the times). A patent was good for a period of fourteen years, subject to a seven-year extension if a special board (later just the Commissioner) could be convinced that the inventor had not yet realized an appropriate profit on his invention.

While the number of applications grew steadily, the proportion of applications that proved patentable varied enormously, depending on the prevailing administrative philosophy within the Patent Office. The Patent Act of 1836 remained fundamental, although it was frequently amended. The first change in the requirements was made less than a year later: Two sets of drawings were now to accompany each application, one to remain in the official files, the other to be attached to the patent itself and returned to the patentee. The impetus behind this had been a fire in December 1836, which destroyed everything in the Patent Office. In the 1840s, '50s, and '60s, there were nearly two dozen amendments to the law, including the following: New and original designs *became patentable in*

1842 (the first design patent went to George Bruce of New York City for a typeface); the term during which the patent remained in effect was increased from fourteen years to seventeen in 1861, while the troublesome extension clause was repealed; and, likewise in 1861, discrimination in the matter of fees was eliminated (a telling commentary on those times). The next statutory milestone was the passage by Congress of the Act of July 8, 1870, which consolidated the modifications of the past thirty-four years and finally rescinded the model requirement. On its own volition, however, the Patent Office continued to require models for another decade; but even after it eliminated this requirement once and for all, models still continued to be submitted.

Why Models?

by Kendall J. Dood

President Andrew Jackson signed into law the bill that established a new patent system in the United States on July 4, 1836. The more substantive features of that bill, the ones essentially defining patents and patentable inventions, remain in effect to this day. The law of 1836 also included provisions relating to the formalities of the patenting process, and among these was the requirement that patent applicants deposit models with the Patent Office whenever an invention could be so represented. That provision remained in effect for thirty-four years as a statutory requirement, and ten more as a rule of the Patent Office.

The models submitted during those years were used for the disclosure of inventions, not just to the Patent Office staff but also to other inventors, to patent agents, and to the public as well. The same statute that required the submission of models also required applicants to describe in writing what they had invented, and to provide illustrative drawings. These two other forms of disclosure, written and graphic, were not only the sole legally effective ones—becoming integral parts of the patent grant itself when the patent was issued—but, in certain respects, they must have been more informative than the model. The written description, together with the drawings, was required to be sufficiently detailed and complete to enable anyone skilled in the technological realm to which it pertained to make or use the invention, as well as to point out precisely the novelty and the utility of the device. No such requirements were imposed on the model. Hence the model was not only redundant as a disclosure, it was, in terms of required information content, inferior to the corresponding specification and drawings.

Why, then, were models required in the first place?

The answer to this question is not as obvious as one might hope. It is possible, of course, that the requirement was simply the result of all the partial reasons in its favor; more likely, though, it reflected some distinct need perceived at the time it was written into law in 1836, a need which only models were believed able to satisfy. Reasons set forth during the actual duration of that requirement were usually based on ad hoc functions the models served in the Patent Office in those years and thus do not necessarily reflect the *original* reasons. Both the report of the Congressional committee which prepared the bill that became the Patent Act of 1836 and the record of the Congressional debates leading up to its passage are strangely silent about reasons for the model requirement. In view of this absence of official explanation in the earliest records, it may never be possible to ascertain the underlying factors or their relative importance. Nevertheless, enough relevant information from the period is available to suggest at least a probable explanation.

The basic patent law in effect previously, the Patent Act of 1793, gave the Superintendent of Patents the

Examiners at work in the US Patent Office, 1869
Harper's Weekly, July 10, 1869

authority to require models *if* he deemed them necessary. While models were at first required only "if the machine be complex," in the years immediately preceding 1836 the Patent Office began calling for them in every possible case. Apparently, as the validity of more and more patent claims was contested in courts of law, models were assuming increasing importance as evidence of the exact nature of the patented invention. The Act of 1793 left the *first* test of patentability to the courts; the Patent Office was concerned only that applications meet certain formal requirements, and before a patent was issued there was no examination of the novelty of an invention nor of the sufficiency with which it was disclosed. The substance of an application was officially "examined" only when a patentee attempted to exercise his patent rights by suing an alleged infringer, and when the alleged infringer, in turn, countered that the patent was invalid either because the invention was not new or

because the description was inaccurate or insufficient.

The Patent Act of 1793 and the Act of 1836 were alike in that both based the *ultimate* test of patentability on the newness of an invention, and on a description that both clearly distinguished what was new from what was old and also enabled anyone skilled in the pertinent arts to reproduce the invention. Between 1793 and 1836 the test was waged not in the Patent Office itself but rather in a court of law; if the judge or jury found that a patent failed to satisfy the requirements of novelty and adequate disclosure, it was held to be invalid and inoperative as a legal instrument. Since no examination of these questions had been made *before* the patent was issued, the invalidity defense often succeeded.

An invalid patent was not necessarily fatal to an inventor's rights to his invention, however. Even if the patent claims were ruled to be lacking in novelty, the invention itself might still be novel and therefore still patentable. The problem in such a case was simply that the applicant had not described his invention in terms sufficient to distinguish it from prior inventions. In cases where the description was found to be either inaccurate or inadequate the patentee had the right to apply for a corrected or "reissue" patent, which would then supersede the original patent in every respect. Here, a model of the invention, filed with the original application, might play a unique and valuable role, for often it was only the model that was able to establish for the Patent Office what the original invention *was*.

As the number of patents increased, the number of successful challenges to their validity increased too, and this, in turn, led to an increase in the number of reissue patents. Thus, the value of the original patent *documents*—as a definitive statement of the subject-matter protected by the patent—became secondary to that of the *model*. Models were closely consulted not only by the Patent Office staff in reissue cases, but also by patentees, alleged infringers, and the courts in infringement cases, and in general by inventors, prospective manufacturers, and licensees anxious to avoid infringing or otherwise interfering with the legitimate rights of patentees.

By 1828, models had become so important to the proper functioning of the patent system that the Office began requiring them not just in cases of complex machines but whenever "it is believed that the nature of the machine will be more clearly shown by [a model] than by drawings alone." Then, in 1834, the rules were changed again so that a model was required "in all cases." Thus when the new statutes were drafted two years later, inventors were already accustomed to submitting models with their applications—and that was

undoubtedly why the Congressional committee responsible for the Patent Act of 1836 did not feel constrained to comment on the model requirement. This merely made statutory what had already become established practice.

It is, however, one of the many ironies of the patent model requirement that the factors which served to make this an accepted part of the patent system before 1836 could not have justified the requirement afterwards. With the statutory change, models were no longer needed as alternatives to inadequate or defective disclosures in patent specifications, because the Act of 1836 established an examining board; individual examiners were to make certain that these were accurate and complete and that the patent claims described subject-matter that was truly patentable. Defective applications were still received after 1836, of course, and in such instances the model provided some helpful supplementary information for the patent examiner. But information present in the model was not permitted to serve in lieu of information lacking in the specification and drawings. At most, it might enable the examiner to suggest ways in which the specification and drawings might be acceptably revised. The submission of a proper specification and drawings was an absolute prerequisite to the granting of a patent. Hence models were no longer needed, as they had been before, to clarify vague or inaccurate patent disclosures.

Quite a different need for models had appeared evident as the statutory changes were impending, a need embodied in the concerns of lawmakers and patent officials regarding the extent to which the new system would be accepted by inventors. The changes were indeed radical. Besides stipulating that the specification

and drawings had to be examined for accuracy and completeness, the law also required that, if the Patent Office determined that the invention lacked novelty, the application be rejected. This provision was particularly worrisome for two reasons.

Two years before, the Patent Office had begun unofficially "advising" applicants of the novelty of their inventions. An upshot of this experiment was the discovery that inventors were, in the words of one official, "exceedingly tenacious in regard to the originality of their inventions, and would not yield without the most conclusive testimony." Inventors, it seems, were not convinced that mere government functionaries were competent to judge the patentability of new inventions, and were thus not anxious to see that authority transferred from the courts. As a contemporary commentator on patent law put it,

The administrative officer . . . should be invested with authority to reject the application in case he deems the claims to be plainly immoral or illegal, or not intelligibly or distinctly specified, [but] I think it is very difficult, if not absolutely impracticable on experiment, to carry preliminary adjudication beyond these bounds.

The second reason for concern about the examination provision of the new law was that it appeared to place the Patent Office in a role *vis à vis* the inventor of both adversary and judge. Where, before, the question of novelty was debated by the patentee and the alleged infringer and then decided by presumably disinterested third parties in courts

of law, under the new act the applicant simply presented his invention to the Patent Office and hoped for a favorable decision. Though there were provisions for appeal from examiners' decisions, the primary concern was that the authority of the Patent Office to deny patents would be viewed by inventors as "discretionary." The commentator just quoted also observed:

It will readily occur that this discretionary authority, whereby functionaries of the government are empowered by act as they may think expedient, is precisely the most odious, and most liable to abuse, that can be granted. . . . The matter becomes one of solicitation instead of being one of right.

Because of inventors' adherence to the notion that patents were a matter of "right" rather than "solicitation," the proponents of the new patent bill were anxious to have the system appear democratic and fair. To this end, they stressed that the power to reject for lack of novelty should be exercised only in the most obvious cases and only on the basis of irrefutable proofs. At the time, the Patent Office's collection of more than 7,000 models appeared to offer such proofs, not only because the models had been accepted as such in the past when the corresponding specification and drawings were so often inadequate, but also because they were a kind of tangible evidence that would not require the interpretation of juries or the "construction" of judges, as the documentary descriptions of inventions often did. Hence the model collection, originally used to help patentees establish the true nature of their inventions, was now to be enlarged and used to help the Patent Office convince applicants whose inventions were not new that what they had was not patentable.

This explanation for the model requirement, however, itself raises another question: If the models were indeed considered by the authors of the Patent Act of 1836 to be essential as proofs of prior invention, why was this not one of the explanations given after 1836, which were based on actual practice? To this question there are at least two clear answers. The first involves another of the ironies of the history of the model requirement: Six months after the new law was passed, a fire destroyed the Patent Office and all the models in it. Thus models were not used as proofs, simply because at first there were none extant. As for the second answer, it is perhaps more significant since it also explains why there was no lasting reluctance among inventors to accept a system that entailed the risk of rejections for lack of novelty, even without the sort of proof the models had been expected to provide. Inventors, it appears, soon realized that the examination process instituted by the new law was not only administered competently and fairly, but that it also resulted in patents much more valuable for being less apt to contain the defects that had plagued them before. Ironically again, the examination requirement which, before the law was passed, had seemed to make the models necessary, in practice made them unnecessary.

In view of the ready acceptance of the examination process instituted by the Act of 1836, it is in some respects surprising that the model requirement remained in the laws as long as it did. Even though models were redundant as disclosures of inventions and as proofs of prior invention, the mere fact that they arrived with every application and were subsequently put on public display tended to promote their use as alternatives to the written descriptions and drawings. There was also the matter of sheer habit—the models were used as disclosures simply because they were there and because they had been used for that purpose in the past, even though the specifications and drawings of patents issued after 1836 were more authoritative than before. This casual usage of the models not only tended to perpetuate the requirement longer than could likely have been justified on the basis of need, it also tended to delay the introduction and distribution of *printed copies* of patent specifications and drawings.

It is no coincidence that the routine printing of patents began only when the model requirement was dropped from the statutes, long after other nations had begun printing patents. The effects of the failure of the U.S. government to disseminate patent information more widely had been felt soon after the system was reinvigorated in 1836. Official, detailed information on patented inventions was readily accessible only in *one* place, at the Patent Office, either from the models or from the original specifications and drawings kept there as part of the public record. While the Act of 1836 authorized the commissioner to prepare copies of patent documents upon request, this was done at the requestor's expense (by hand, of course), and the expense was prohibitive except where copies of specific patents were needed for business or legal purposes. For anyone merely seeking some idea of the "state of the art," there was at first little alternative but to hire someone to make a search of the Patent Office records, or to make a trip to Washington for a personal search.

Attempts to rectify this situation started as early as 1843, when Commissioner Henry Ellsworth began to include a claim from each patent issued during the year in his annual report to Congress. Extra copies of the commissioner's reports were distributed throughout the land, and soon these also included a drawing from each patent and an abstract of the specification. But this only served to whet the appetites of mechanics and artisans, patentees and would-be patentees. While technical journals such as *Scientific American* contributed to the dissemination of details concerning patented inventions, it was becoming clear that only publication of the *whole patent document* would satisfy the demand. It took so much money to house, tend, and display the Patent Office's burgeoning model collection that funds were not available for printing patents. Eventually, however, it seemed obvious that the ready accessibility of complete, inexpensive, printed copies of all patents would render the model collection superfluous— at least insofar as the dissemination of information to inventors was concerned. After the model requirement was dropped from the statutes, the Patent Office began printing copies of all patents as they were issued, and subsequently old patents were put into print as well. In 1880 the model requirement was dispensed with.

By 1893 it had become quite clear—again, insofar as the dissemination of information to inventors was concerned—that there was no need to keep the model collection on exhibit in downtown Washington, D.C., and that year the process got under way of removing the patent models from the Patent Office permanently.

The most pressing exigencies behind the repeal of the model requirement from the statutes and then from the rules of the Patent Office undoubtedly were the ever-increasing expenses of storage, display, and maintenance, and the competing demands for space. The demand for printed patents was just as important, if not more so, as was the fact that inventions in new fields of technology lent themselves less well to the meaningful representation in the form of models than had the simpler machines and mechanical devices of earlier years. What ultimately made feasible the repeal of the requirement and disposal of the model collection was the simple truth that models did not perform as irrefutable proofs of prior invention, essential as "conclusive testimony" in cases of rejection. The culminating irony is that they probably never had.

Kendall J. Dood *is a patent classifier for the U.S. Patent and Trademark Office in Crystal City, Virginia. This essay is a condensed version of a two-part article published in the* Journal of the Patent Office Society.

The Patent Models on Display

Douglas E. Evelyn

When Congress stipulated in the Patent Act of 1836 that models be submitted with patent applications, it had more in mind than mere determinations of "novelty, originality, and utility." It recognized, too, the virtues of *exhibiting* models as a means of educating the public, stimulating creativity and enterprise, and inspiring a sense of America's destiny. In his report of April 28, 1836, Senator John Ruggles of Maine, Chairman of the Select Committee on the Patent Office, noted that since the War of 1812, America had "become all at once a manufacturing, as well as an agricultural and commercial nation." The accomplishments registered in just a few decades, Ruggles claimed, "would have taken Europe a century. . . ." Patent models had a role as exhibitable evidence of these achievements.

In the first years of the patent system, only a few models were collected. The first Patent Board, which consisted of three part-time examiners, Secretary of State Thomas Jefferson, Attorney General Edmund Randolph and Secretary of War Henry Knox, granted only fifty-five patents in its first two years. The next year, 1793, the mandatory requirement that models accompany all applications was eliminated. At that point, the patent process became little more than a registration system with no tests for novelty of inventions and hence no protection to the patentees. Models came in randomly and were displayed in the Patent Office which from 1810–1836 was housed in cramped quarters at the former Blodgett's Hotel which also accommodated the City and Federal Post Offices.

This state of ineffective chaos persisted until the convening of the Ruggles Committee in 1836. Its findings resulted in the enactment of two laws on July 4, 1836. The first called for a rigorous evaluation process to be followed in the granting of patents under the direction of a Commissioner of Patents. The second provided for a new, fireproof Patent Office Building to be constructed on a site bounded by F and G Streets, between 7th and 9th, where Pierre L'Enfant had once proposed locating a great National Church.

Henry L. Ellsworth was appointed as the first Commissioner, and the architectural plans submitted by Ithiel Town and William P. Elliot were approved by Congress. To run the project, President Jackson appointed the architect Robert Mills, an expert on fireproof construction. Soon after work began on the new site, a fire broke out in Blodgett's Hotel on December 15, 1836, demolishing the Patent Office's quarters and destroying most of its records, drawings, and books along with about 7,000 models. Senator Ruggles issued a report on the fire, itemizing the losses and recommending that 3,000 of the most important models be replaced and reiterating that the Patent Office's new quarters be a fireproof building "with the necessary cases and furniture." Congress appropriated

$100,000 to replace the destroyed models and records, and construction of the new building continued while the office took up temporary residence in the Old City Hall.

The first wing of the Patent Office Building was completed in 1840, at a cost of $415,000, of which somewhat more than one-quarter came out of the Patent Fund, accumulated from the fees paid by applicants. On December 18, 1840, Patent Commissioner Ellsworth issued the plan of exhibition for the new building. His statement was laced with patriotic rhetoric, describing a "National Gallery" on the upper floor that would "remain a perpetual exhibition of the progress and improvement of the arts in the United States." As suggested in the Ruggles Report of 1836, it was to include fine art, specimens of "useful and elegant fabrics," possibly "a cabinet of interesting minerals" and "a collection of Indian curiosities and antiquities," and "such other objects of interest as might conveniently and properly be placed under the superintendence of the [Patent] Commissioner." Ellsworth made no mention of patent models, however, and, in fact, when it came to contributing to this display of "the genius and industry of the nation," the Patent Office was initially at a disadvantage: It had been almost exactly four years

since the calamitous fire at Blodgett's Hotel, few of the destroyed patent models had been replaced despite his efforts, and the new ones that had accumulated since 1836— less than 2,000—could not fill much of the grand gallery.

With neither enough patent models nor "articles of industry" to compete effectively for space, Ellsworth had to sit by as the Patent Office became a kind of polyglot "Cabinet of Curiosities." The historic collections of the State Department were ensconced on the upper floor, along with Indian artifacts and portraits from the War Department and the miscellaneous holdings of the National Institute for the Promotion of Science, an organization which sought to turn James Smithson's bequest to the purposes of a museum. After 1842 a glut of specimens brought back by the United States Exploring Expedition which had ranged southern latitudes for four years nearly filled the National Gallery.

The shortage of patent models did not persist, however. By the end of the 1840s, the collection had grown to its former size, the building was attracting as many as 10,000 visitors per month, and most of the models had to be held in storage. Complaints about "usurpation" of space became a standard litany both with the Patent Office staff and its clients. Fortunately, relief came in 1849 when Congress consolidated a variety of government bureaus including the Patent Office under the aegis of the

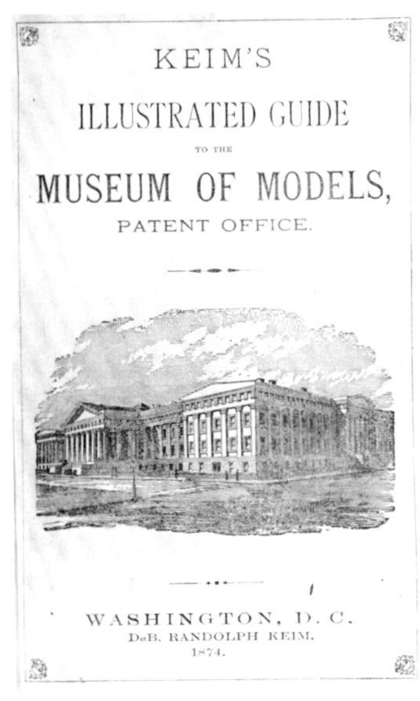

Illustrated Guide to the Museum of Models Patent Office Building
by DeB., Randolph Keim
Washington, D.C., 1874.
Photograph courtesy of
    National Portrait Gallery,
    Smithsonian Institution,
    Washington, D.C.

Museum of United States Patent Office
United States Magazine, 1856
Photograph courtesy of
    National Portrait Gallery,
    Smithsonian Institutions,
    Washington, D.C.

Home Department (today's Interior) and appropriated funds for enlarging the Patent Office building.

An east wing facing 7th Street was completed in 1852 and a west wing facing Ninth in 1856. While territorial disputes with other Interior Department bureaus concerned with Indians, lands, and pensions continued, for a time, at least, there was sufficient space to mount a comprehensive display of patent models. Guidebooks made much of the models as a tourist attraction and visitors from home and abroad recorded their impressions of this display which, according to *United States Magazine* for October 1856 comprised more than 25,000 "specimens of ingenuity and skill . . . judiciously arranged by subject."

As the cumulative number of patents in 1856 was substantially less than 25,000, that display would have included several thousand models that had accompanied applications which were *rejected*. Rejected models would have served instructive purposes, of course, but keeping them on display was a practice that hastened the day when the Patent Office would be awash in models.

The halcyon days when the Office could display almost all the models it had, "judiciously arranged," were first interrupted in 1861 when the Civil War brought new tenants to the building, and its wide corridors and vast galleries, now including a north wing, were converted into improvised barracks and hospital wards. In 1865 the third-floor north gallery served as the magnificent setting for Lincoln's second inaugural ball. About the "sort of fascinating sight" of the Patent Office as a hospital, Walt Whitman wrote:

Two of the immense apartments are filled with high and ponderous glass cases, crowded with models in miniature of every kind of utensil, machine, or invention it ever entered into the mind of man to conceive. . . . Between these cases are lateral openings, perhaps eight feet wide and quite deep, and in these are placed the sick, besides a great long double row of them up and down through the middle of the hall.

After the war—as the number of patented models topped 100,000—the north wing, which had been completed in 1860, became available for display. It was the final side of a continuous quadrangular gallery a quarter-mile long, making a full circuit of the upper floor. Most of the miscellaneous collections displayed in the original (south) wing had been transferred to the Smithsonian Institution and it became possible to devote this enormous gallery almost exclusively to models, patented and unpatented, ranged in row upon row of glass cases. By the late 1860s, the number of patented models was growing at the rate of more than 13,000 annually, and the gallery was being progressively double-decked to keep ahead of the influx.

The potential number of models seemed infinite, but space to display them definitely was not. Whereas the model requirement had once appeared to make sense in terms of stimulating creativity and fostering national pride, the problems of accommodating the quantities of models were increasing. As Commissioner Samuel Sparks Fisher noted in 1869:

It must soon become a serious question to determine what disposition is to be made of the models. In a vast number of cases no such illustration is required; the inventions are so simple that they can readily be understood by good drawings. Models soon become broken or inoperative, and in some cases they have been altered surreptitiously, so as to become false witnesses.

We may take just pride in our national museum of the mechanic arts, but it is questionable whether this museum can be allowed to grow at the rate of five thousand square feet per annum.

A remedy . . . will probably eventually be found in dispensing with all models, except when, in the discretion of the Commissioner, such mode of illustration is absolutely necessary.

Heeding Fisher's advice, Congress dropped the requirement for models when it legislated a new Patent Act in 1870, but allowing the Commissioner to request them when necessary. Fisher's successor, M. D. Leggett, considered models generally useful "for full and clear understanding," and he used his discretionary authority to call for them routinely. Moreover, many inventors ignored the new law and continued to submit models with their applications.

By the year of the Centennial, the fortieth anniversary of

Sleeping Bunks of the First Rhode Island Regiment at the Patent Office, Washington
Harper's Weekly, June 1, 1861. Photograph courtesy of National Portrait Gallery, Smithsonian Institution, Washington, D.C.

the Patent Act of 1836, the model collection was consuming space even more insatiably than Fisher had anticipated. There were about 175,000 models, jammed into 362 cases, from 400 to 1,600 in each one. In the last available upper-deck space there was room for twenty-four additional cases, not sufficient even for two years at the rate models were then accumulating. Little semblance of a "judicious arrangement" remained. Only token relief was planned—selective disposal of models that had accompanied rejected applications. That would scarcely suffice, but, when major relief did come, it was not planned at all.

On September 22, 1877, a fire destroyed over 76,000 models housed in the two newest wings of the building, west and north. Between 1879 and 1885 the damaged portions of the building were rebuilt, along with the upper floor of the south wing, which was in poor condition. When reconstruction was finished and the surviving models reinstalled, the concept of the quadrangular gallery was retained, although in the south hall models were confined to decks above a range of offices. The pressure of demands for office space and storage for paper records was re-

lentless, however, and within a few years displacement of the model exhibitions had begun. A loan collection was installed in the Smithsonian's National Museum in conjunction with the celebration of the centennial of the first Patent Act, and most of this was subsequently transferred for continuing exhibition. In 1892 the Patent Office occupied only about two-fifths of its own building, and models had been moved into passageways as galleries were converted to offices or file rooms. After surveying the scene, a reporter lamented, "Many of the model cases are so closely crowded together that the doors cannot be opened, and an examination of their contents is impossible." Obviously, models that were not even accessible did not need to stay in one of the government's primary office buildings, and in 1893 the Commissioner reported that more than half of the remaining 155,000 had been moved to rented storage space, where they were *totally* inaccessible.

In 1908, Congress first sought to get rid of the models. After the Smithsonian selected 1,061 associated with famous inventors, efforts to auction off those remaining resulted in the transfer of only 3,000 others. The rest, packed in thousands of oak crates, went back into

storage until 1925 when Congress established a commission to develop a plan for giving those that were historically important to the Smithsonian and other institutions and disposing of all the others by any means feasible. An accurate inventory was and remained an impossible task. The great majority ended up in private hands, relatively few in historical collections.

The various entrepreneurs who promoted models recognized their appeal to "Americana" buffs, and the museum curators to whose charge they fell understood that they were historically important. It apparently never occurred to anyone that they might *all* be important as documents of nineteenth-century American invention.

Occasionally someone voiced dismay about the way patent models were exploited as curios. Yet, in view of how they were neglected by the government during the years they were shuffled about, it seems fortunate that so many do survive in private hands—far better to have been dispersed to a thousand mantlepieces than rotting in leaky livery stables, or gone altogether. As always, hard numbers are largely

guesswork, but it is conceivable that as many as half of the patent models may survive today.

What has been lost without any hope of retrieval is the ambience of the grand quadrangular Model Gallery, an entire floor of a building that was once both sizable and stunningly appointed. Images survive only in old photos and woodcuts printed on the deteriorating pages of periodicals such as *Harper's Weekly*. Contemporary exhibitions of patent models cannot possibly recreate the experience. The sad reality is that the patent models will never again be seen the way they were—ranged in hundreds upon hundreds of display cases around the "national museum of the mechanic arts" through the upper reaches of a classic building that stood where Pierre L'Enfant had envisioned a great National Church.

Douglas E. Evelyn *is Deputy Director of the National Museum of American History, Smithsonian Institution. He was formerly Deputy Director of the National Portrait Gallery, which occupies half of the Old Patent Office Building in Washington, D.C.*

Technological Aspirations

by George Nelson

What most of us see when confronted by a collection of patent models is a mass of fragments of a toylike world, whose dimensions have been determined by a government regulation to fit within a cube twelve inches on a side. This rule generally held whether the invention submitted was a clothespin or a railroad locomotive. There probably never was a period when miniatures have failed to enchant the viewer, so it would be normal for most of us, with no aspirations to invention, to see in century-old models submitted for patents the expressions of a dollhouse world and to respond to the instant, universal appeal of the miniature. Because the models are old, we can also see in them artifacts created by a vanished culture, there for all to examine for clues to what life must have been like way back then. The fact that so many thousands of these little objects, now precious and instructive mementos, have been mauled, neglected, lost in dusty files and display cases, exposed to fire, and auctioned off by speculators, adds a certain pathos to the examination.

Scanning the history of these models, one is struck by its resemblance to the tale of the sorcerer's apprentice, in which a boy is left in charge of his master's magic perpetual motion machine, pouring out its product without interruption, only to realize that he had forgotten the incantation needed to turn it off. In the 1830s the Patent Office turned on its almost-perpetual motion machine by requiring that inventors submit models in addition to drawings and written descriptions. By 1880, when the requirement was dropped, the Patent Office was overflowing with models. And still the models kept coming in. After the Columbian Exposition in 1893 more than 155,000 remained, and in 1908 the Congress finally addressed itself to the problem of their disposal. A hardy breed, this Lilliputian assemblage of crystallized ideas.

Cultural artifacts have a way of swinging wildly in the scales of public esteem. Rome's Colosseum was viewed during the Middle Ages not as a relic of the glories of the Empire but as a handy quarry whose stones were already cut and ready to use again. At the other extreme we have the example of the home of the late Elvis Presley transformed into a shrine which is incidentally the core of a fat business fed by the visits of millions of pilgrims.

As a people, we are very *immediate:* we want what we want right now, and once tired of it we forget it immediately. In the nineteenth century, our now-neglected patent models were one of the major tourist magnets of Washington, D.C. Important voices were raised to inform the public that the models were there "to elevate our national character," a rather astonishing proposition when you think about it, and a very revealing one as well. The Patent Office Building was set down on the site reserved in L'Enfant's plan for a great National Church. One might surmise that in the absence of such a monument, an exhibition dedicated to 100,000 patent models might be an appropriate alternative. The universal optimism that permeated the land then is hard to evoke in these days of doubt. It is necessary to remind ourselves that this was indeed God's country, the biggest and best thing that ever happened. If the streets were not exactly paved with gold, at least they were getting paved, and no matter how modest our performance in this life, our children were going to do better. All we had to do was wait through a century of saltwater taffy, softball, cinder-clogged parlor cars, and ice-powered refrigerators until television and computer games were finally invented.

Artifacts are remarkably clear transmitting devices, for while words can mislead and even lie, artifacts cannot. They are by their nature "on the level," and their messages, for those able to decode them, are always truthful. When we look at the model of Abe Lincoln's contraption for getting riverboats off sandbars (although it was never constructed), we can see in it an entire vanished world in action: men heaving and perspiring with ropes, pulleys and winches, bodies in the shallow water pushing and pulling; crews and passengers looking over the side of the mired craft, listing in its sandy trap; we can see people on the shore, too, in tight-waisted blouses and big hats. Those were the days when even near-disasters had a holiday quality. I picked this as a sample of a "readable" artifact in part, I am sure, because the picture of Lincoln as an inventor is hard to resist, but also because it is so obvious that the images it evokes are *not* in the invention, but in popular prints of the time, or old books, or even a film about life on the Mississippi. Because of the peculiar nature of perception, what we "see" is not necessarily what is there, but rather a series of parallel constructions made up of relevant images.

Examples of this very human blending of fact and imagination can be found everywhere. There is the marvelous story of Archimedes and his cry of "Eureka!" which has been echoing in the corridors of memory for 2,000 years. The Tyrant of Syracuse had been given a gold crown, and with the suspicion normal in an experienced ruler he wondered if it was really made of gold or alloyed with some inferior metal and therefore called in the mathematician Archimedes for an answer. Archimedes knew that pure gold would weigh more than an alloy. He needed to determine the exact volume of the crown so that an equivalent mass of gold could be weighed. It was while taking a bath that the solution came to him. He knew that every time he got into the tub the water level rose. This time, however, *he saw in this fact the displacement principle:* The volume displaced was always exactly the same as the volume of his body. Putting the crown in a vessel full of water, all he had to do was catch the overflow in a measuring cup,

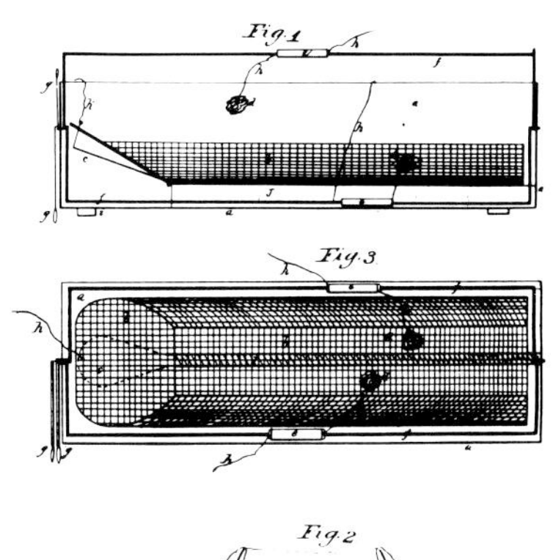

Improvement in Electric Baths
Mark W. House
Cleveland, Ohio
February 18, 1862
Patent No. 34,425
(see also page 58)

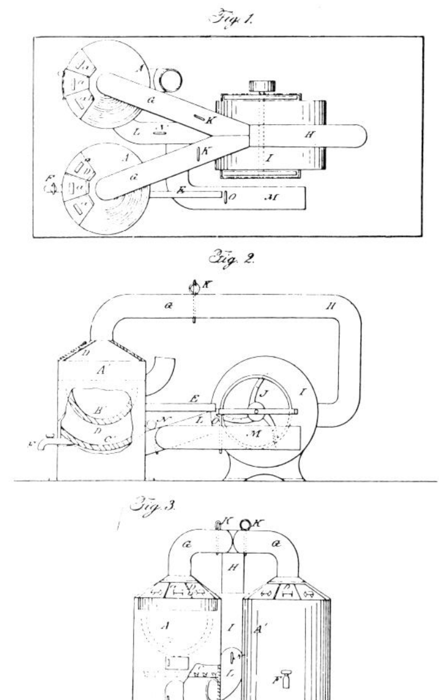

Improved Apparatus for Rendering Tallow
John J. Eckel and Isaac S. Schuyler
New York, New York
August 14, 1866
Patent No. 57,104
(see also page 86)

which gave him the precise volume of the crown. Duplicating this volume with gold, he weighed it and got his answer. Had he merely seen the water go up without envisioning the necessary connection, nothing would have happened.

All major inventions must follow the pattern of Archimedes' great discovery. Whether the procedure is contrived or, like the discovery of vulcanized rubber, accidental, the pattern does not seem to vary greatly. What does vary is the scale and quality of the idea. Archimedes achieved an extraordinary insight into Nature's workings, applicable from then on to any number of situations. It is unclear whether there really are such things as "breakthroughs" in science and technology. More often changes occur gradually as a multitude of small improvements and innovations accumulate. Certainly, in the period of the patent models the tens of thousands of inventors, both amateur and professional, almost invariably homed in on small localized problems: an improvement to a cutting machine, a new variation on a

steam pump, a fastener, an ingenious piece of multi-use furniture. The American temperament, then as now, was one adapted to a sharp focus on practical problems. A wide-ranging intellectual curiosity was not in the repertory of the average citizen, nor was there any reason why it should be. The society was dynamic and unstable; life was filled with survival efforts mixed with a new striving toward the goals of fame and fortune which had become possible with the creation of an egalitarian society. The urge of the anonymous multitudes of immigrants, once they had established a foothold, was to move up to a respectable standard of living as rapidly as possible. The combination of personal freedom, an immense continental territory, and a rising standard of living released an enormous amount of human energy, largely directed to attainable goals like making money—and assuring a better life for future generations.

The picture presented by the models bears all of this out. It was generally assumed that the best place to make things in large quantities was the factory, that industrial methods had to displace the handicrafts,

and, since production methods are endlessly open to improvement, the notion that change always equalled progress was accepted without question. Even today, notwithstanding the virtual disappearance of such unquestioning optimism, a solid majority of our population continues to believe that "new" always means "better," as every *new, improved* package of laundry detergents, headache remedies, breakfast food, and cosmetics can testify.

One can only guess regarding such matters, but it seems reasonable to assume that the act of invention is not a simple matter of a single individual perceiving a need for a particular device, but rather of several individuals becoming aware of the possibility for a certain kind of device through the efforts of others to solve a particular problem. Then it becomes possible to think of better ways of doing the same thing. In a way it is like the process of evolution in nature, where changes in the genetic pool interact with an environment, with natural selection determining which types have the greatest survival potential.

The *content* of inventions, as

we can see in the patent models, is always different from the *process of inventing*. The latter is the outcome of unpredictable behavior on the part of individuals; the content is almost always socially determined. Nobody, in a world where war was unknown, could invent an MX missile: there would be no way of even thinking of such a device. The Greeks knew that steam generated pressure, but they had no interest in labor-saving inventions since there was an ample supply of slaves. Such tight linkages between what societies are and what they do make it possible to use inventions as documents that can reveal something of the nature of the society in question.

The projects in the domestic area, which apparently interested a large number of inventors, have by nature a certain fuzziness at the edges: A piece of furniture does not have the need for extreme precision required in a power tool or a steam engine. Moreover, there is a very noticeable difference in quality between models submitted by the stars in the inventing business, names like John Ericsson, Matthias Baldwin, George Corliss, and George Westinghouse, whose models were always precisely

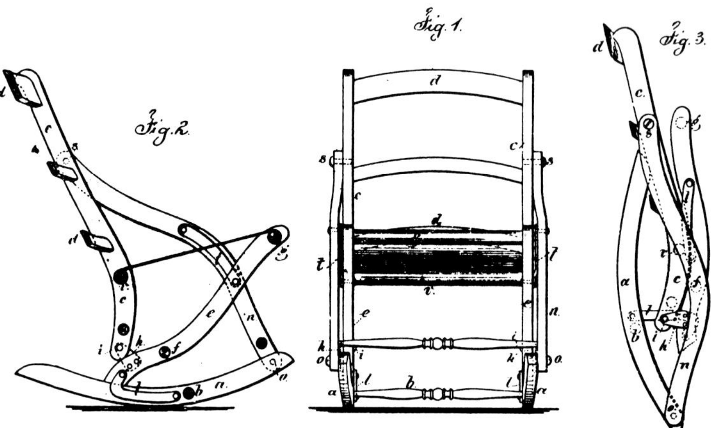

Improvement in Folding Rocking Chairs
Ephraim Tucker
Worcester, Massachusetts
June 18, 1878
Patent No. 205,015
(see also page 57)

made of quality materials, and the majority of the submissions, which were generally of a much lower standard. One sees exactly the same kind of discrepancy in any contemporary architectural or design competition, where the winners not only display better thinking, but also presentations on a higher level of quality as well. An inventor's work can be evaluated in the same manner. There is another side to this coin, however, which is that peak performance only reaches a significant level where there is a large participating group, whether consumers or an audience. In activities as disparate as baseball, ballet, bullfighting, and the ancient Greek drama, the common element is a large and highly involved audience that knows in detail the difference between top performance and mediocrity. It is entirely possible that the tremendous upsurge in the entire U.S. economy during the nineteenth century and its great success in capturing export markets was in some way connected with the vast numbers of aspiring amateurs who kept flooding the Patent Office with their ideas.

What is really fascinating about many of these pieces is the ideas themselves, some of which sweep the observer past the limits of credibility. One combination unit begins as a box and opens up to include a bed, a table, a chair, a bureau, and a commode. Also ready for service in all beds, folding or rigid, there is a "handy bedbug trap" which, even though it may not have worked to protect the sleepers, does at least indicate something about general living conditions. As always in such matters, logic does not necessarily produce the right answers. Granting that multi-purpose furniture makes sense in crowded living spaces, why would anyone produce a *folding* rocking chair? Any such piece, no matter how cleverly designed, has considerable bulk when folded, hardly solving the problem of overcrowded space. It is also difficult to imagine the family that would rock, then fold, then rock again.

It is always tempting to stretch partial explanations to cover a total situation. If much housing was unduly cramped, then it is reasonable to expect the appearance of multi-purpose furniture. However, it is just possible that many of the strange hybrids concocted for patent approval were created by inventors who were simply having a ball thinking up problems calling for ingenuity. Such challenges may have been a major motivating force in the entire process of inventing. One of the oddest assumptions of our supposedly rational society is that no one ever does anything just for the hell of it, that sheer delight is never a motivating element in work, that beyond money and a rise in social status there are no possible reasons for working at anything. This utterly barren view of existence, which now seems to dominate thinking at all levels, is surely going to do us in if we do not sometime wake up to a realization that life is meant to be lived, not sold or rented in conveniently packaged fragments.

It may be the most naive kind of wishful thinking, but what I get out of these models is not so much that they are now obsolete in a world where technology has moved on to much more powerful and less visible tools, and have thus passed on to survive as curious samples of folk art. This is indeed the case, and of course as cultural artifacts they are as valuable as Indian arrowheads and Neanderthal flints, with much to tell us about the forms of a life that has changed beyond recognition, and not always for the better. No, the thing that comes through for me is the still undiluted magic of the creative effort at any level, the excitement of pursuing an absorbing problem, the feeling that one's efforts are going to benefit one's fellow man, the simple joy of taking an idea into one's own hands and giving it proper form.

This is the point at which professional skill and amateur fumbling come out as essentially the same thing, and it may be one of the oldest existing manifestations of our common humanity, as old as the touching little Venus of Cyprus or the superb painted walls of Lascaux and Altamira—only this time around it is patent models, and they should be regarded with the same delight and respect.

George Nelson, *designer, teacher, and writer was trained as an architect. Among his published books are* Problems of Design *and* How to See: Visual Adventures in the World God Never Made.

Plates

a.
**Colter or Joiner Supporter for
   Plows**
James Oliver
South Bend, Indiana
May 8, 1877
Patent No. 190,510
SI

b.
Cutter Bar
Obed Hussey
Baltimore, Maryland
August 7, 1847
Patent No. 5,227
SI

The Farm

Charles Newbold patented the first cast-iron plow in 1797; it failed commercially not because (as legend has it) farmers feared it would "poison the soil," but simply because it cost too much in relation to its limited life-span. Others continued to improve upon cast-iron plows, especially after Jethro Wood began producing patented plows around 1817 with readily interchangeable components. Wood's "Eagle" plow was popular for many years, though it was ultimately not as successful as John Deere's steel-edged plow (which scoured so well it became famous as the "singing plow").

Steel plows were expensive, however, and one of James Oliver's signal contributions was his patented method of hardening cast iron. Oliver began experimenting with "chilled" iron in 1865, gradually improving his metallurgical techniques for another decade, and finally patenting the adjustable plow shown here. By the time of Oliver's death in 1908, the Oliver Chilled Iron Plow Works was producing 200,000 plows annually.

Like many models of agricultural implements, the 1877 Oliver patent model was a perfect miniature, with the patented elements set in the context of the entire device. Obed Hussey's model, by contrast, exemplifies a different approach: Only the patented component is shown out of context. Hussey had patented the first reaper in December 1833, six months before Cyrus McCormick patented his version. Ultimately McCormick's proved the superior machine in all ways except for the design of the slotted finger bars of the cutter; that was Hussey's invention, and it is the feature that *still* predominates in cutting machines. Hussey and McCormick fought a monumental series of patent battles, and much discussion ensued over the question of who was "the inventor" of the reaper. As is usually the case, the best machine was the result of the efforts of several inventors refining and improving a basic design.

c.
Harvester
G. T. Coolman and C. M. Young
Corry, Pennsylvania
February 15, 1870
Patent No. 99,850
SI

d.
Changing Harvester From Reaper to Mower
Robert Beans
Johnsville, Pennsylvania
August 28, 1855
Patent No. 13,504
SI

25

a.

Apparatus for Drying Grain

Thomas F. Rowland, James
  Stephens, and William H.
  Mason
Brooklyn, New York
May 22, 1855
Patent No. 12,922
SI

b.

**Improvement in Cotton Seed
  Planters**

J. A. Stewart, John Stewart,
  and William H. Prister
Franklin, Kentucky
July 1, 1856
Patent No. 15,260
SI

c.

Cultivator

R. M. Brooks
Greenville, Georgia
June 26, 1860
Patent No. 28,829
SI

d.
Flail
Patent Office Replica, 1892
SI

a.

b.

In 1790, about 90 percent of the American population were farmers, and 60 percent remained farmers as late as 1860. It is understandable, then, why farm implements and apparatus for ancillary operations—such as flailing and drying hay—received intensive attention. Inventions like Oliver's and Hussey's remain a part of our living history, yet far more typical are the thousands of prosaic inventions noteworthy only insofar as they suggest how little novelty was actually required to satisfy the Patent Office when it was in one of its "liberal" phases. There were apparently any number of ways to make barbed wire, for example, and fully a decade before A. C. Decker received his barbed-wire patent in 1884, Joseph Glidden had patented the basic improvement that had enabled large-scale production of this fencing for the treeless plains.

As for Denzil Phillips's sheep-feeding rack, the primary features—that it could be disassembled and folded up, and that it prevented animals from "wasting their feed"— were certainly desirable. Yet a functionally equivalent apparatus (altered sufficiently to avoid infringing Phillips's patent) could have been put together by practically any farmer. Whether this rack or the variant forms of barbed wire were ever commercially produced remains unknown.

a.
Barb Fence
Alexander C. Decker
Bushnell, Illinois
June 3, 1884
Patent No. 299,916
DM

b.
Barb Fence
[No number]
PC

c.
Sheep-Feeding Rack
Denzil Phillips
Cochranton, Pennsylvania
February 2, 1886
Patent No. 335,384
PC

d.
Hay-Fork
Samuel G. Simpson
Mill Creek, Pennsylvania
November 30, 1869
Patent No. 97,449
SI

The Household

Because patents tend to be associated with the history of technology, and because "technology" calls to mind "manufacturers' goods"—factory engines, shop machinery, commercial transportation—it is easy to forget how common it was and is for the impetus to invent to be turned toward household items. Yet, of the surviving patent models, a very substantial proportion relate to patents on devices for cooling, cooking, heating, lighting, cleaning, sewing, washing, drying, and to all manner of furniture and cabinetry. Peter Welsh writes that "this large number of patents for household appliances reflects an increased standard of living in a new nation where comfort and convenience gradually emerged as middle-class prerequisites newly compatible with the older gospel of 'Poor Richard.'"

Ironically, these models which depict improvements to stoves, washing machines, lighting devices and the like, are most compelling as evidence of the drudgery of daily life in the nineteenth century. Clothes and dishes still have to be washed; roaches and dust are still bothersome intruders; but electric vacuum cleaners, dishwashers and similar equipment make it difficult to imagine how endless these now paltry chores must have been. Few of those countless inventions, although often creative, could have made any real difference.

a.
Improvement in Kitchen-Cabinets
William H. Stewart
Butler, Missouri
November 12, 1878
Patent No. 209,936
PC

b.

c.

d.

b.
Combined Dinner-Pail and Lantern
John L. Rickard
Salem, Ohio
June 24, 1881
Patent No. 245,402
JF

c.
Improvement in Baskets
Horace C. Jones
Dowagiac, Michigan
February 10, 1874
Patent No. 147,328
PC

d.
Packing-Box for Eggs
Ignatz Karel
Blue Earth City, Minnesota
May 27, 1879
Patent No. 215,993
PC

a.
Refrigerator
William Ferris; Philip Garrett;
  James Megratten
Wilmington, Delaware
April 6, 1858
Patent No. 19,837
PC

b.
**Improvement in Combined Water
  and Liquor Cooler**
William A. Jones
Erie, Pennsylvania
December 17, 1872
Patent No. 133,990
PC

c.
**Improvement in Portable Ice-
  Houses**
John Lippitt
Wilmington, North Carolina
August 14, 1877
Patent No. 194,097
PC

Cooling

As urbanization put more and more Americans beyond ready access to fresh milk, meat, and vegetables, the ice box became increasingly popular. Naturally the demand for ice grew, especially in the cities of the South. Consumption in New Orleans, for example, increased from less than four hundred tons a year in the late 1820s to more than twenty times that much in the 1840s, by which time ice had become a staple New England export for domestic and world markets. Eventually, writes Daniel Boorstin, "the word 'ice' in various combinations had provided more new compounds than perhaps any word in the English language."

One such combination was "ice house," a place from which ice was retailed to consumers who used it in various kinds of coolers. Ice was the most perishable of commodities, of course, and it was to this problem that John Lippitt addressed himself. In his patent specification Lippitt noted that ice shipped to the South was ordinarily put in barrels or boxes to be sent to the "interior country" by rail or water, and there it was again transferred to ice houses, a procedure that entailed "very serious waste." Lippitt's scheme involved packing "portable ice cars aboard ship," then moving them to their final destination with no further transferral, and selling the ice directly from the cars. Here was a remarkably early example of containerized transport.

33

a.

a.
**Improvement in Atmospheric and
    Vacuum Egg-Beaters**
Henry E. Marchand
Pittsburgh, Pennsylvania
April 9, 1879
Patent No. 214,936
PC

b.
Apparatus for Making Coffee
John Denley and Thomas H.
    Heberling
Warsaw, Illinois
August 3, 1858
Patent No. 21,066
PC

c.
Improvement in Culinary Vessels
C. A. Johnson
Des Moines, Iowa
July 7, 1868
Patent No. 79,574
PC

b.

c.

d.
**Improvement in Holders for
    Preserve-Cans**
Mrs. C. Dewey How
Cincinnati, Ohio
August 3, 1875
Patent No. 166,380
JF

d.

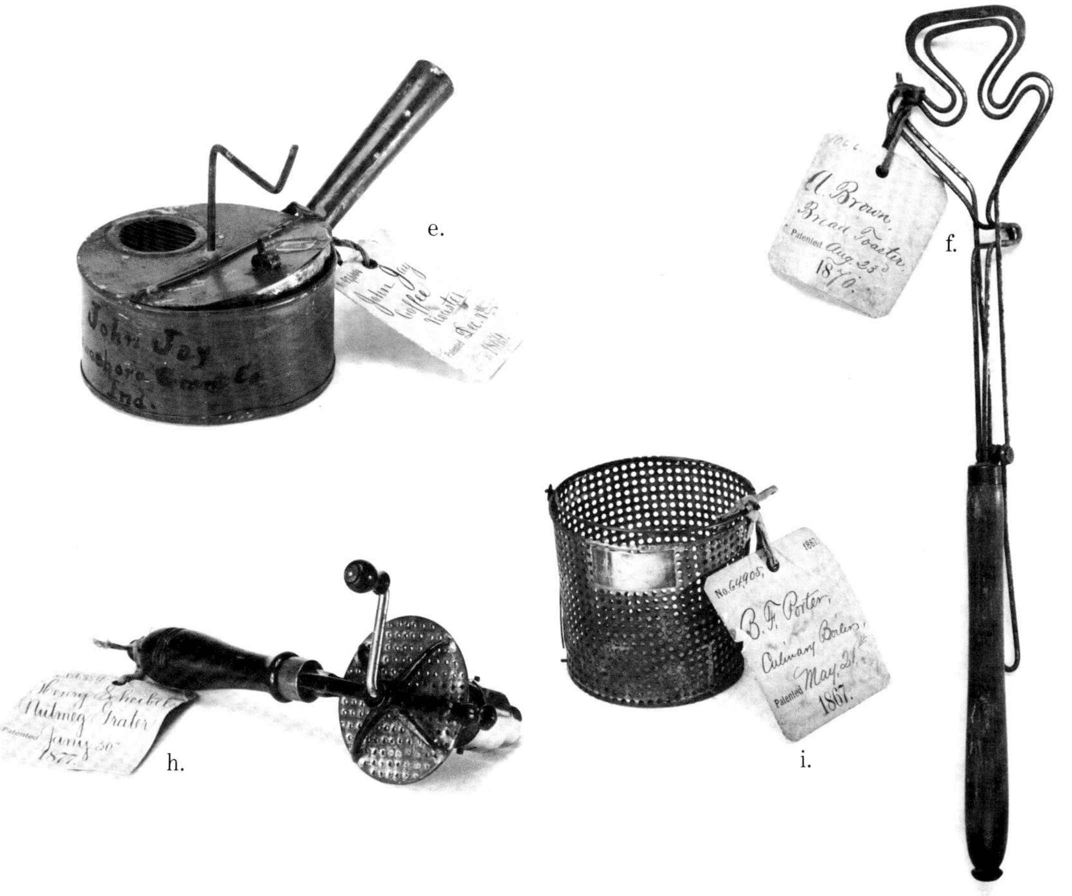

e.

f.

g.

h.

i.

Kitchen Utensils

Here, by and large, are the most prosaic sorts of inventions; yet none of them should be dismissed as trivial. The kitchen, after all, was the workplace of a very substantial proportion of the population, and the models reflect the pervasiveness of a set of values centering on convenience, simplicity, and economy of both cost and function. These values were essentially codified in 1869, in *The American Woman's Home*, a book on domestic management by the sisters Catherine Beecher and Harriet Beecher Stowe. They aimed to demonstrate that, with "scientific" planning, a middle-class housewife could manage her home without the servants that seemed necessary in the larger homes of the upper class.

The models tend to be prosaic, yet especially appealing because so often they appear to have been homemade, perhaps assembled on the inventor's kitchen table. Most were probably never commercially produced, and yet these types of gadgets are somehow still present in most kitchens. The Read "spider" had an integral hollow handle "adapted to the grasp of the hand"; Johnson's culinary vessel was a "combined apparatus for steaming and mashing potatoes and other vegetables, and keeping them warm"; Jay's coffee roaster was "simple," "portable," and "capable of being converted into a device for popping corn"; the Denley-Heberling coffee pot was a percolator that could "be made from common tin plate"; the Marchand egg-beater was "cheap, neat, and handy"; the Schneider nutmeg grater did its job "without the least waste"; the Brown bread toaster could also be used as a pancake turner.

e.
Coffee-Roaster
John Jay
Jonesborough, Indiana
December 7, 1869
Patent No. 97,644
PC

f.
Bread-Toaster
Alanson Brown
Rochester, New York
August 23, 1870
Patent No. 106,654
PC

g.
Improvement in Spiders
Josiah M. Read
Boston, Massachusetts
June 10, 1879
Patent No. 216,346
PC

h.
Improvement in Nutmeg-Graters
Henry Scheibel
Bridgeport, Connecticut
January 30, 1877
Patent No. 186,884
PC

i.
Improved Culinary Boiler
Benjamin F. Porter
Manchester, New Hampshire
May 21, 1867
Patent No. 64,905
PC

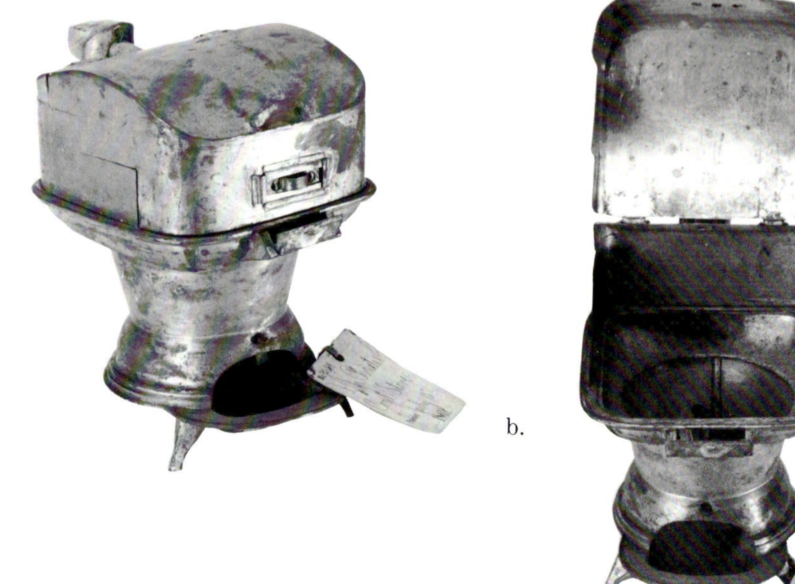

b.

a.

a.
Oven
John P. Hayes
Philadelphia, Pennsylvania
July 31, 1855
Patent No. 13,375
SI

b.
Coal Stove
Isaac J. Baxter
Peekskill, New York
July 13, 1869
Patent No. 92,511
PC

Heating

In the first decades of the nineteenth century, writes Eliphalet Nott's biographer, America "was still a nation of hearths." As iron castings became more readily available, however, and especially as the technology of burning "stone coal" (anthracite) was developed, it was becoming a nation of stoves. A turning point here was Nott's base-burning "Saracenic Stove," patented in 1826, with its "rotary grate." Nott introduced many subsequent improvements, and it is said that by the late 1830s the Nott stove "was as familiar a household object as a Windsor chair." Nott's manufacturing company ultimately failed, but iron stoves continued to provide a staple product for entrepreneurs such as Jordan Mott, proprietor of the Mott Iron Works.

Models for stoves and fireplaces abound, which suggests that heating was a constant problem and that its solution was always in need of further refinement. By the 1850s, inventions pertaining to heating and cooking tended to be accessories such as warming shelves, dampers, and scuttles. As with all household devices, there were the inevitable combinations, such as the Baxter stove—which was convertible into a heater for tailor's sadirons—and the combination stovepipe, damper, and shelves by Clinton and Lavinia. Most of the models were of homemade tinware; some, such as the Hayes oven model, were professionally cast.

c.
Improvement in Stove-Pipe Drums
Henry Meyer
Richmond, Indiana
September 8, 1868
Patent No. 82,019
SI

d.
Improvement in Stove-Pipe Shelves
George B. Johnston and
  Judson Allen
St. Charles, Missouri
October 24, 1876
Patent No. 183,681
PC

e.
**Improvement in Combined Stove-
  Pipe Dampers and Shelves**
Edward H. Clinton and
  William H. Lavinia
Chicago, Illinois
May 23, 1871
Patent No. 115,025
PC

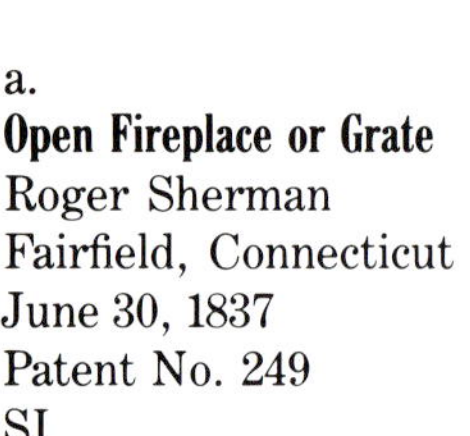

a.

Open Fireplace or Grate
Roger Sherman
Fairfield, Connecticut
June 30, 1837
Patent No. 249
SI

b.

Improvement/Fireplace
David Hayes
Chicago, Illinois
December 28, 1875
Patent No. 171,565
SI

c.

**Constructing the Flues of Open
    Fireplaces**
Thomas Whitson
New York, New York
June 14, 1838
Patent No. 784
SI

d.

**Combined Coal-Scuttle and Ash-
    Screen**
A. F. Carling and F. Rockwell
Ellenville, New York
October 31, 1865
Patent No. 50,685
PC

e.

Improved Hot Air Grate
C. B. Loveless
Syracuse, New York
August 5, 1862
Patent No. 36,098
PC

f.

**Portable Steam Radiator for
   Heating Apartments**
J. H. Chester
Cincinnati, Ohio
June 30, 1857
Patent No. 17,666
SI

a.
Improvement in Lanterns
William Westlake
Chicago, Illinois
September 6, 1870
Patent No. 107,140
SI

b. and c.
Lamps for Burning Lard
B. K. Maltby and J. Neal
Rootstown and Middlebury,
 Ohio
May 4, 1842
Patent No. 2,604
SI

d.
Lard Lamp
Dexter H. Chamberlain
Boston, Massachusetts
August 29, 1854
Patent No. 11,633
SI

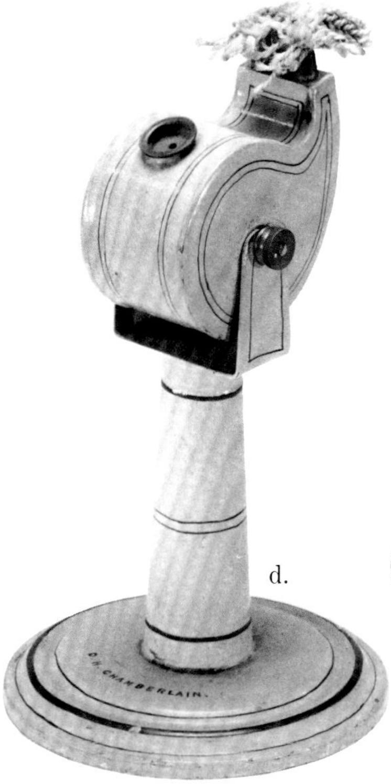

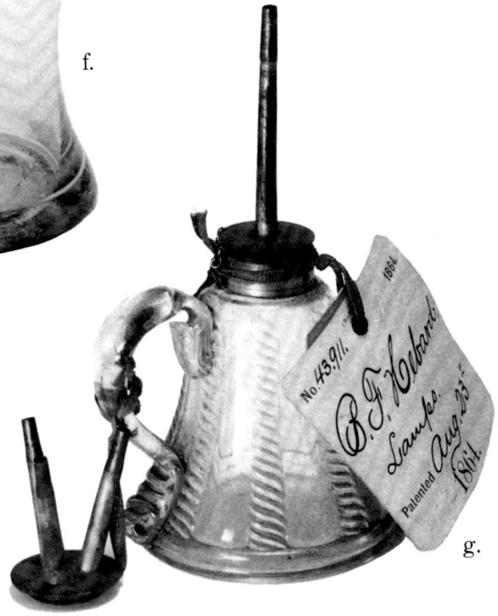

e.
Design for Lamps
W. W. Bacon
October 18, 1843
Design Patent No. 12
SI

f.
Mode of Attaching Extinguishers to Lamps
F. A. Jewett
Abington, Massachusetts
December 11, 1855
Patent No. 13,910
SI

g.
Improvement in Lamps
Benjamin F. Hebard
Dorchester, Massachusetts
August 23, 1864
Patent No. 43,911
SI

h.
Lard Lamp
Jeremiah Senseny
Chambersburg, Pennsylvania
July 15, 1856
Patent No. 15,364
SI

Cleaning

The tasks associated with household cleaning are repetitive and unending; Hannah Arendt, in *The Human Condition*, calls these tasks "labor" as distinguished from "work," which relates to the production of artifacts of some durability. About the "preindustrial" aspects of housework very little is known, but patents provide one of the best sources of information. The tendency has been to pay more attention to a comparatively sophisticated device such as the carpet sweeper, first patented by Melville Bissell in 1876 and subsequently improved in various ways. More is to be learned from such mundane devices as dishcloth holders, soap dishes, and garbage containers. Some forms have survived to the present almost unchanged—how can one improve, for instance, the dustpan? Others can be known only through patent models, since most full-scale tools were used until they fell apart and no actual examples have survived.

a.

a.

Improvement in Garbage-Holders
Rufus Cook
West Newton, Massachusetts
August 13, 1878
Patent No. 207,016
PC

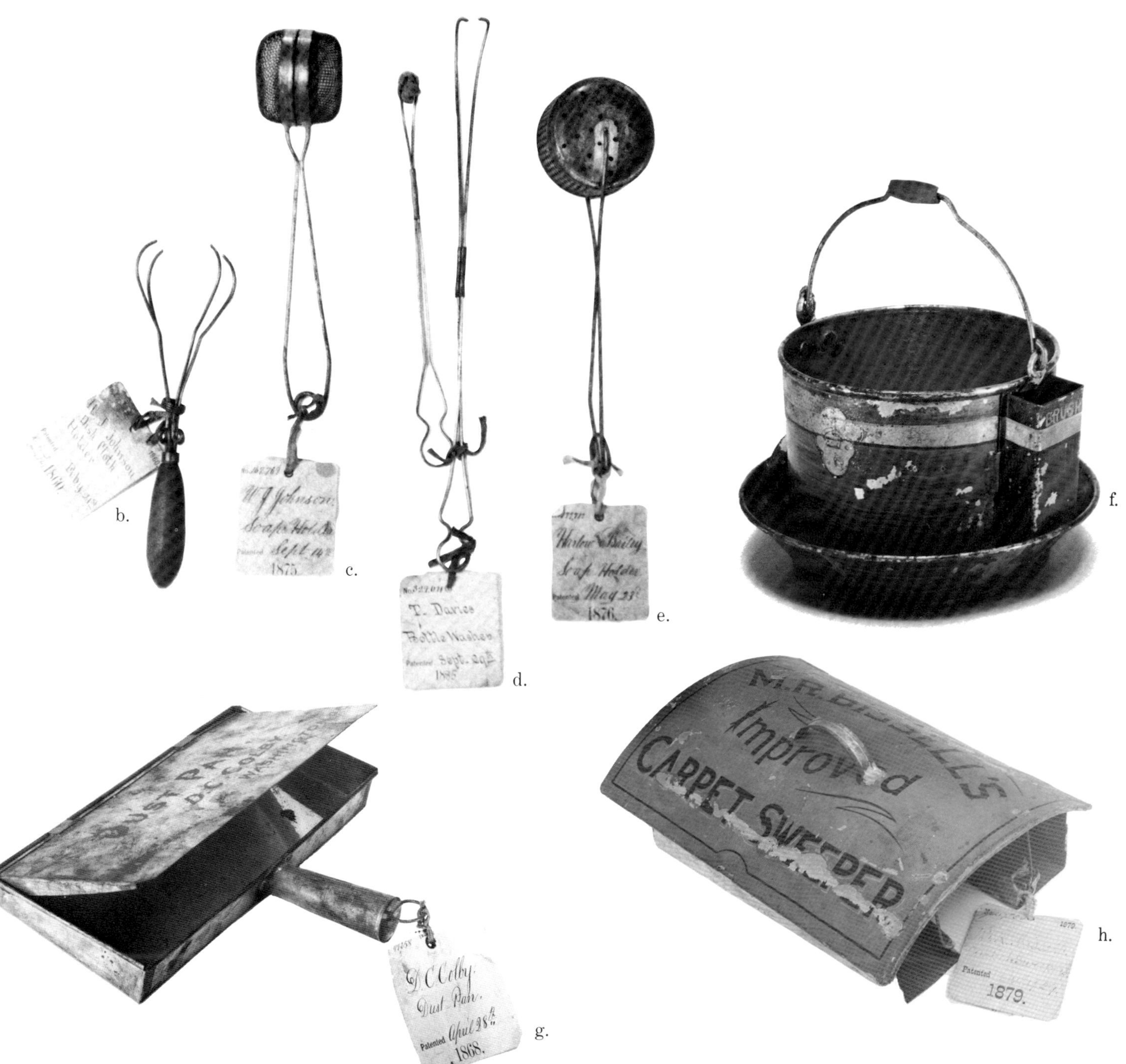

b.
Improved Dish Cloth Holder
W. J. Johnson
Newton Corners,
  Massachusetts
February 21, 1860
Patent No. 27,225
PC

c.
Improvement in Soap-Holders
William J. Johnson
Newton, Massachusetts
September 14, 1875
Patent No. 167,769
PC

d.
Bottle Washer
Thomas Davies
Poughkeepsie, New York
September 29, 1885
Patent No. 327,071
PC

e.
Improvement in Soap-Holders
Samuel Harlow and Horace P.
  Bailey
Plymouth, Massachusetts
May 23, 1876
Paten No. 177,712
PC

f.
Housemaid's Pail
Emma C. Wooster
New York, New York
March 2, 1875
Patent No. 160,500
SI

g.
Dust-Pan
Daniel C. Colby
Washington, D.C.
April 28, 1868
Patent No. 77,358
PC

h.
Carpet Sweeper
Melville R. Bissell
Grand Rapids, Michigan
July 29, 1879
Patent No. 217,854
PC

Sewing Machines

Mechanization in the home in the latter half of the nineteenth century is best exemplified in the sewing machine. Historians have pretty well sorted out the tangled history of the sewing machine's technological development, and have begun to pay attention to its role in fostering an American system of consumption, a phenomenon Neil Harris has called "the drama of consumer desire." The sewing machine as an epitome of American design conventions is well illustrated in the collection of the National Museum of American History, which has more than 700 patent models for sewing machines—a virtually comprehensive survey of this single example. Such curiosities as Clark's dolphin and Perry's horse reflect a desire to domesticate technology, to make objects that are not just useful but pleasing to behold according to the standards of the time.

a.

a.
Sewing Machine
George Hensel
New York, New York
July 12, 1859
Patent No. 24,737
SI

b.
Improvement in Sewing Machine
David W. Clark
Bridgeport, Connecticut
January 19, 1858
Patent No. 19,129
SI

c.
Sewing Machine
Isaac M. Singer
New York, New York
February 6, 1855
Patent No. 12,364
SI

d.
Sewing Machine
G. B. Arnold
New York, New York
May 8, 1860
Patent No. 28,139
SI

e.
Sewing Machine
Elias Howe, Jr.
Brooklyn, New York
January 20, 1857
Patent No. 16,436
SI

f.
Sewing Machine
A. F. Johnson
Boston, Massachusetts
January 13, 1857
Patent No. 16,387
SI

g.
Sewing Machine
James Perry
New York, New York
November 23, 1858
Patent No. 22,148
SI

h.
Sewing Machine
Joseph E. Hendrick
Bristol, Connecticut
October 5, 1858
Patent No. 21,722
SI

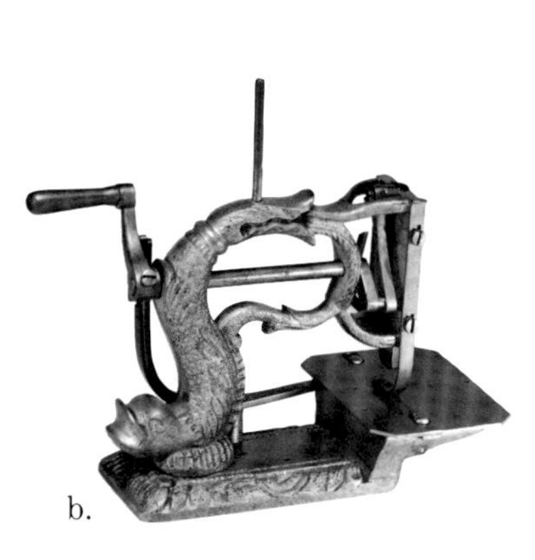

b.

c.

d.

e.

f.

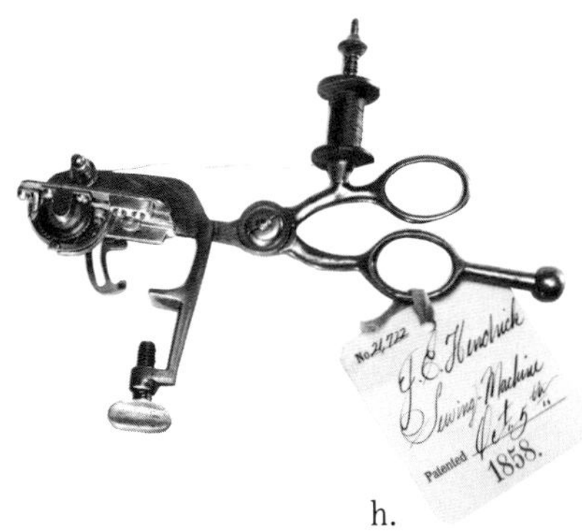

h.

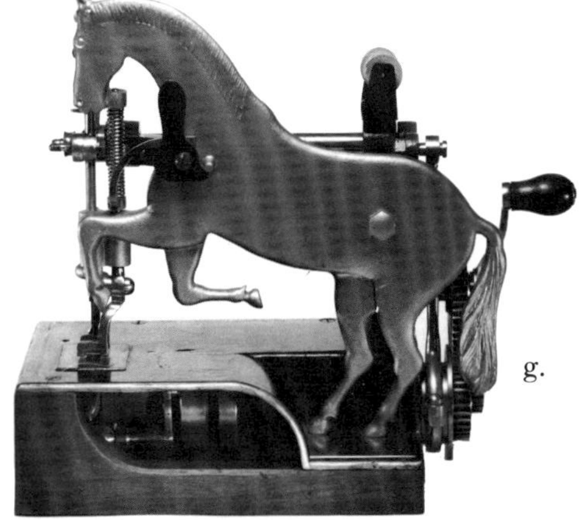

g.

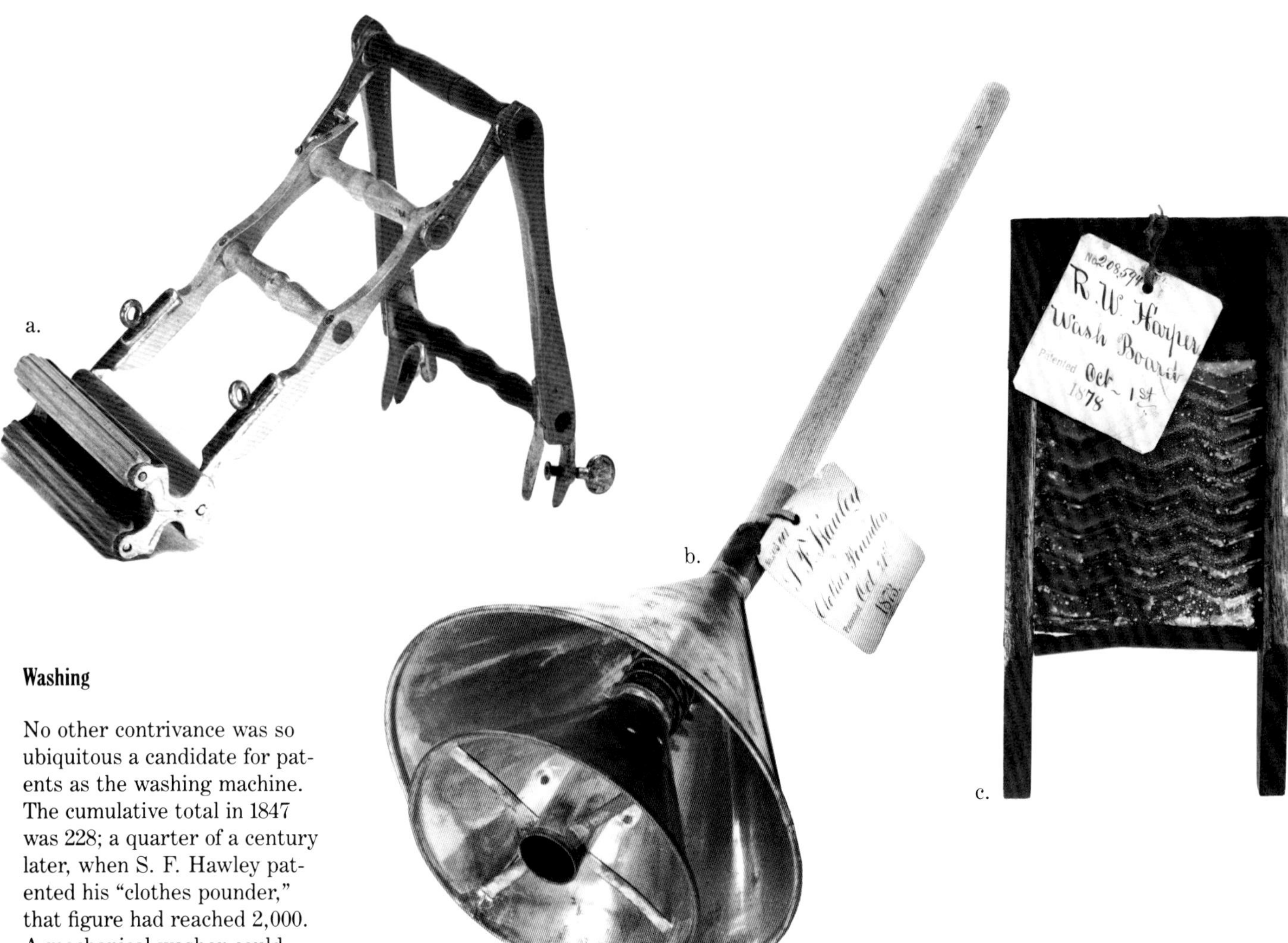

Washing

No other contrivance was so ubiquitous a candidate for patents as the washing machine. The cumulative total in 1847 was 228; a quarter of a century later, when S. F. Hawley patented his "clothes pounder," that figure had reached 2,000. A mechanical washer could assume only so many conceivable forms, so in most cases novelty was marginal—or, if the novelty was apparent, the utility seemed marginal. There were exceptions, however. Among the ranks of washing-machine inventors were a certain number of women, who presumably would have been intimate with the process. One might thus feel that Caroline F. Fleming's "barrel washing machine, the barrel or tub of which is constructed in a horse-shoe form, in transverse section, is corrugated inside, and which is provided with a corrugated roller or rubber of peculiar construction" was something that clearly would have qualified as an "improvement."

a.
Washing Machine
Rachel P. Smith
Chicago, Illinois
May 9, 1882
Patent No. 257,774
SI

b.
Improvement in Clothes Pounders
Samuel F. Hawley
Sandy Hill, New York
October 21, 1873
Patent No. 143,901
PC

c.
Wash-Boards
Robert W. Harper
Evansville, Indiana
October 1, 1878
Patent No. 208,594
SI

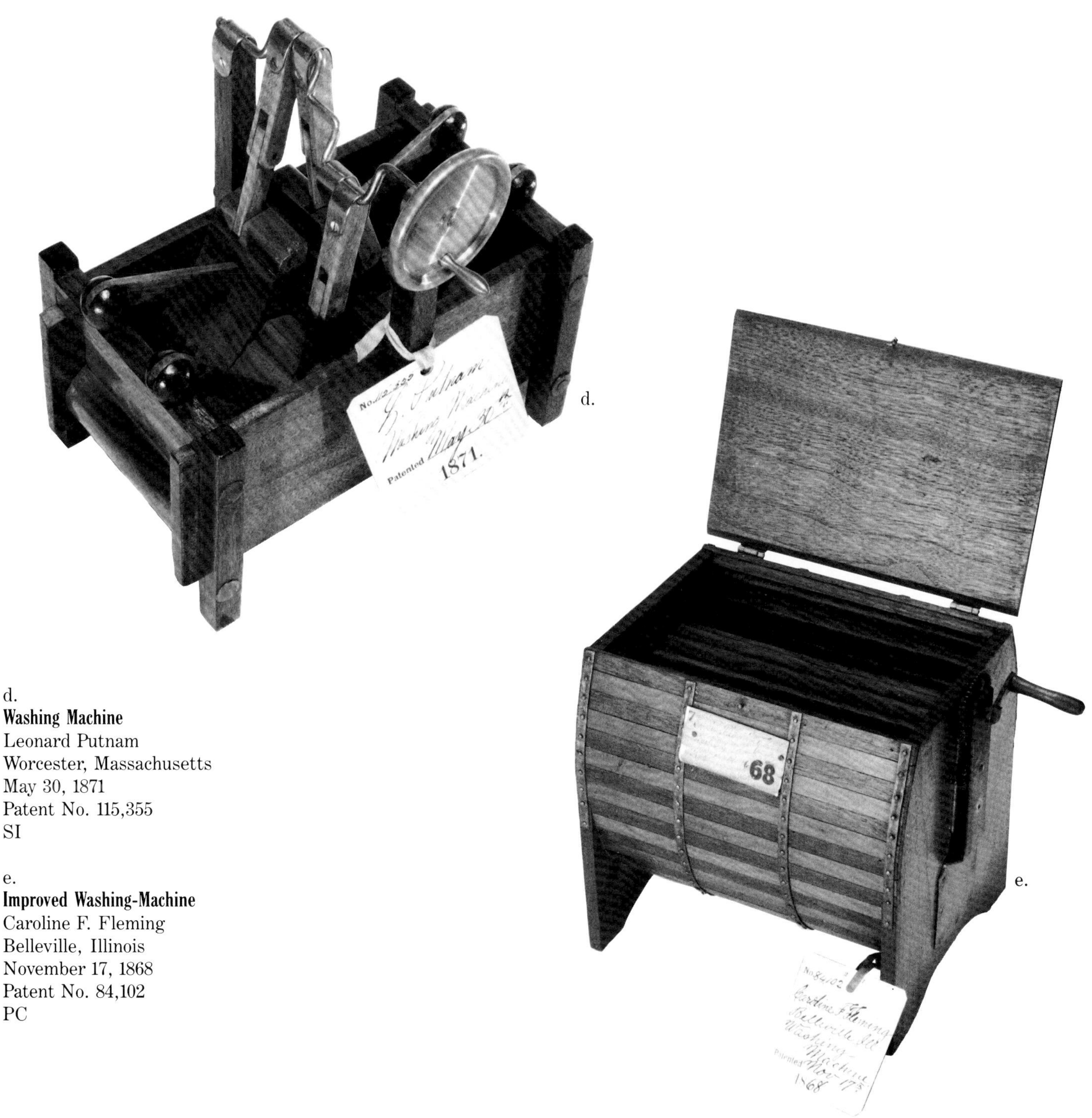

d.
Washing Machine
Leonard Putnam
Worcester, Massachusetts
May 30, 1871
Patent No. 115,355
SI

e.
Improved Washing-Machine
Caroline F. Fleming
Belleville, Illinois
November 17, 1868
Patent No. 84,102
PC

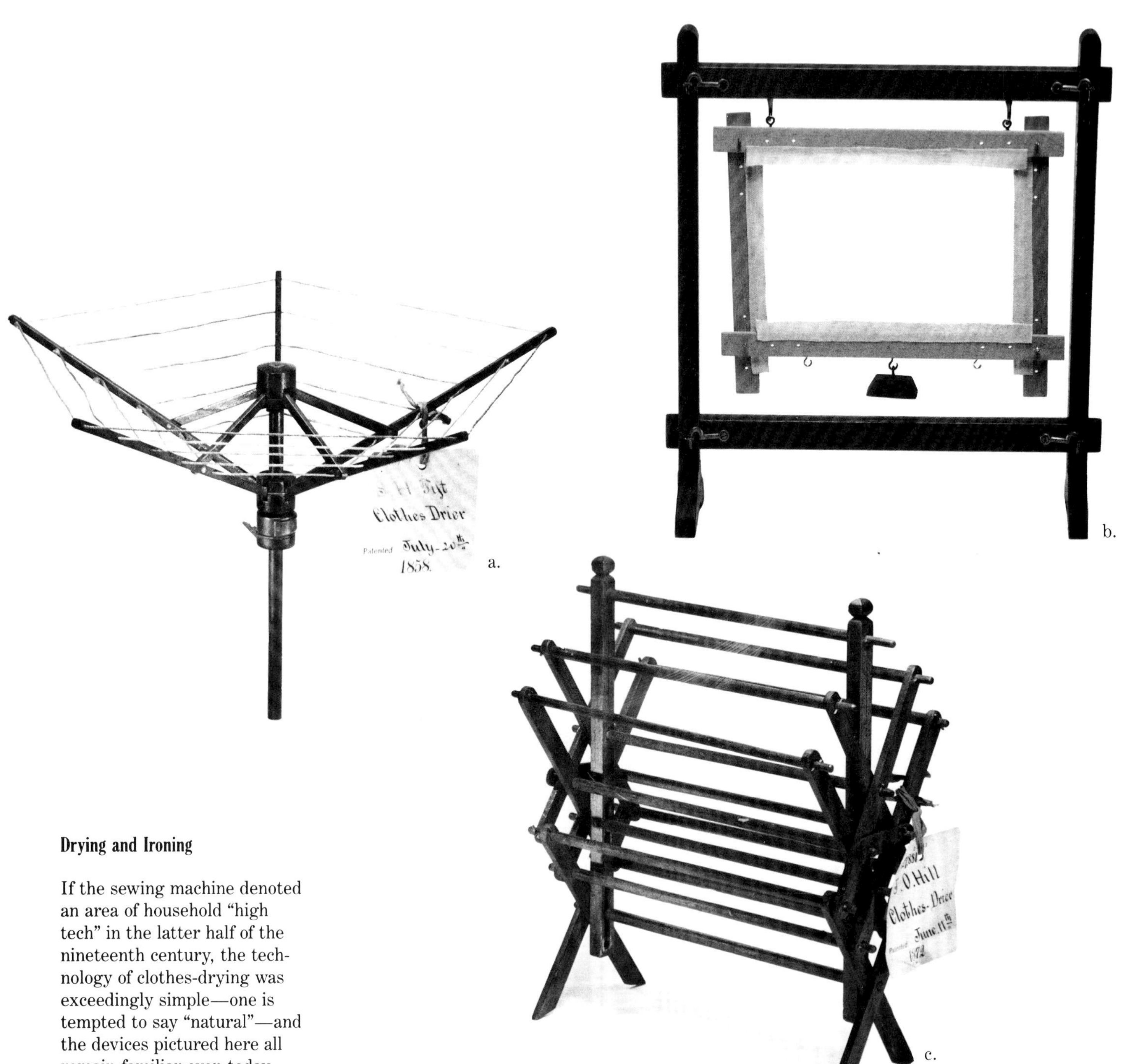

a.

b.

c.

Drying and Ironing

If the sewing machine denoted an area of household "high tech" in the latter half of the nineteenth century, the technology of clothes-drying was exceedingly simple—one is tempted to say "natural"—and the devices pictured here all remain familiar even today. Minor details may vary—one's mind boggles at the number of clothespins patented—but basically all of the pins and clamps secure clothes to ropes or cords. Greater differences can be seen in the more difficult challenge of keeping an iron hot, a problem that has been forgotten since the advent of the electric steam iron.

a.
Clothes Dryer
S. H. Tift
Morrisville, Vermont
July 20, 1858
Patent No. 20,964
SI

b.
Drying Frames for Lace Curtains
S. Short
Cincinnati, Ohio
December 23, 1873
Patent No. 145,912
SI

c.
Clothes Dryer
James O. Hill
Carbondale, Illinois
July 23, 1872
Patent No. 127,881
SI

d.
Heaters for Smoothing Irons
James J. Johnston
Allegheny City, Pennsylvania
January 10, 1854
Patent No. 10,408
DF

e.
Sad Iron Heater
R. Drake
Newark, New Jersey
February 8, 1870
Patent No. 99,541
JF

f.
**Improvement
  in Self-Heating Smoothing-Irons**
William D. Cummings
Washington, Kentucky
October 30, 1856
Patent No. 15,801
PC

g.
Sad Irons
John Lidell
Rockford, Illinois
July 21, 1878
Patent No. 205,656
DF

h.
**Improvement in Folding Table-
  Supports**
Joseph Daly
Troy, New York
June 20, 1871
Patent No. 116,164
PC

i.
Ironing Board
John R. Jackson
Putnamville, Indiana
March 21, 1876
Patent No. 175,106
PC

a.

Home Furnishings

Furniture patents tended to address minor elements of construction, although models generally set the patented feature in full context. The novelty Samuel Reed claimed in his clothes-rack patent was that the two wooden cross-members not only held the hooks in position, but that the hooks in turn held the bars together, the ends being threaded, one right-handed, the other left. While A. G. Mack submitted an elaborate model of a wardrobe, his sole claim to novelty was a revolving hangar-tree "in combination with" fixed shelves and drawers. Elsewhere in his specification, Mack indicated that his wardrobe combined "the convenience of a closet, bureau, and toilet-stand, and, at the same time, occup[ied] but little space." The latter value, compactness, was characteristic of an entire genre of patented furniture, several examples of which appear on the following pages.

"Combined" types of furniture first appear in the patent records with frequency in the 1830s. In addition to an element of evident faddishness

here—and the design challenge of the sort that Rube Goldberg later found so inspirational—a variety of practical reasons account for the popularity of multi-purpose furniture. Clearly the strongest motivating factor in this trend was the steadily growing urbanization of America and the attendant premium which this put on space. After the Civil War, "folding" furniture became increasingly popular, epitomized by the amazingly complex "Wooton Patent Secretary," a desk at which, Rodris Roth has suggested, "much of the business of transforming this country into a powerful industrial nation was accomplished." Most folding furniture, however, was designed not for the office but for the home, with a market stimulated largely by the tenets of the "Domestic Science" move-

ment being fostered by the Beecher sisters. They envisioned homes designed to provide "modes of economising time, labor, and expense by the close packing of conveniences." The language of patent specifications was laden with such phrases as "may be folded up in a suitable recess," and "so as to be wholly out of the way"—characteristics ideal for the "small and economical houses" advocated by the Beecher sisters.

Finally, however, one must not neglect to consider what sorts of challenges were most seductive in an age which sought "a mechanical solution to almost every problem." Peter Welsh points out the prevalence of the added mechanical element, as with the patented bed designed "precisely upon the principle of the windlass."

a.
Improved Combination Wardrobe
A. G. Mack
Rochester, New York
December 14, 1869
Patent No. 97,942
PC

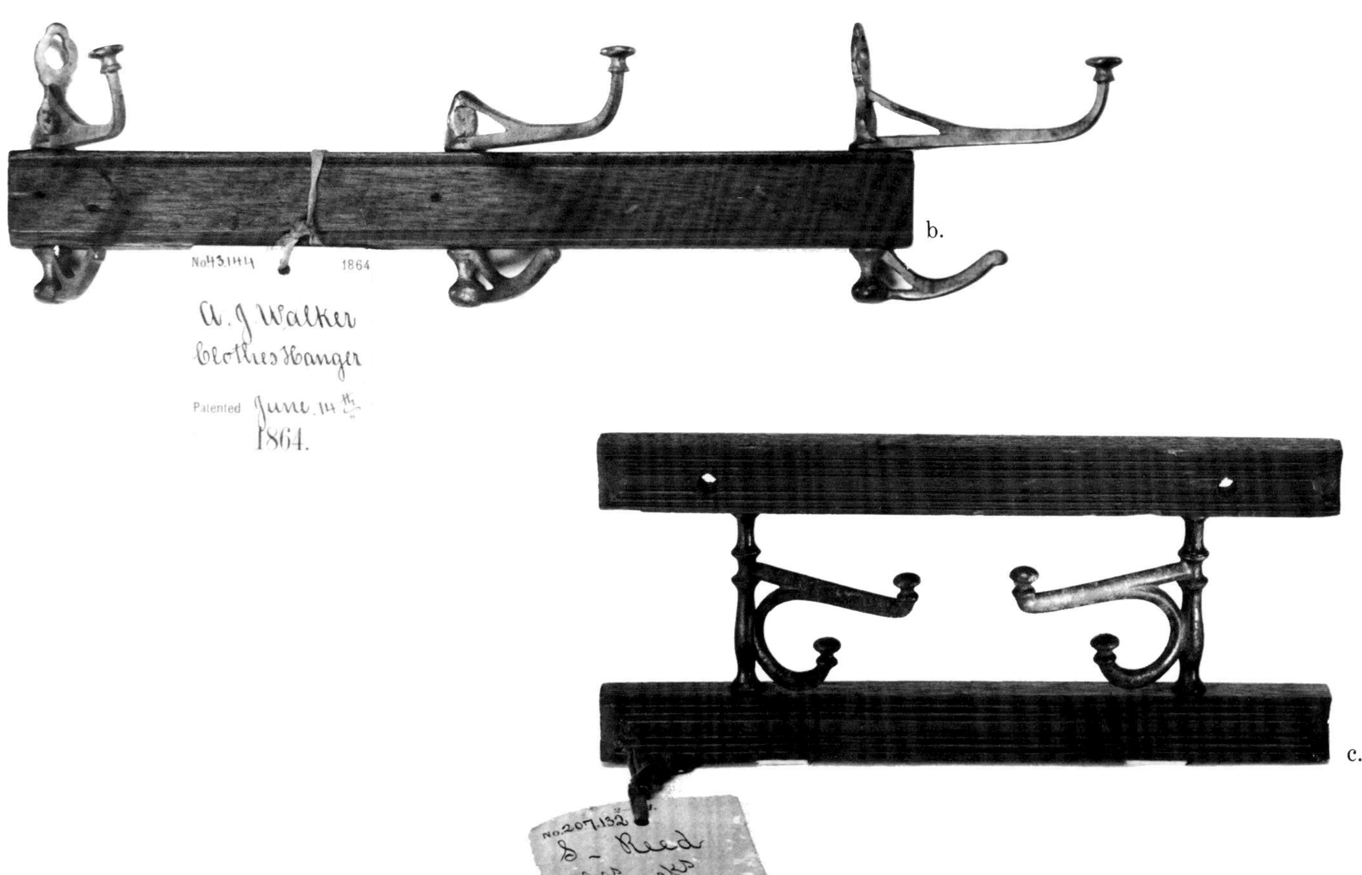

b.

b.
Clothes Hanger
Alexander J. Walker
New York, New York
June 14, 1864
Patent No. 43,144
SI

c.
Improvement in Clothes-Racks
Samuel Reed
Cincinnati, Ohio
August 20, 1878
Patent No. 207,132
PC

a.
Improved Wardrobe & Bedstead
Levi Wing & David Myers
Chicago, Illinois
June 12, 1866
Patent No. 55,568
PC

b.
**Improvement in Combination
  Bedsteads, Chairs and Tables**
Walker Getchell
Bath, Maine
February 15, 1873
Patent No. 142,387
PC, Photograph: Barry Korn

a.

a.
Improvement in Folding Cot-Beds
Hermon W. Ladd, Chelsea,
    Massachusetts
Thomas B. Raymond,
    Stoneham, Massachusetts
and James I. Spencer,
    New York, New York
June 3, 1879
Patent No. 216,194
PC

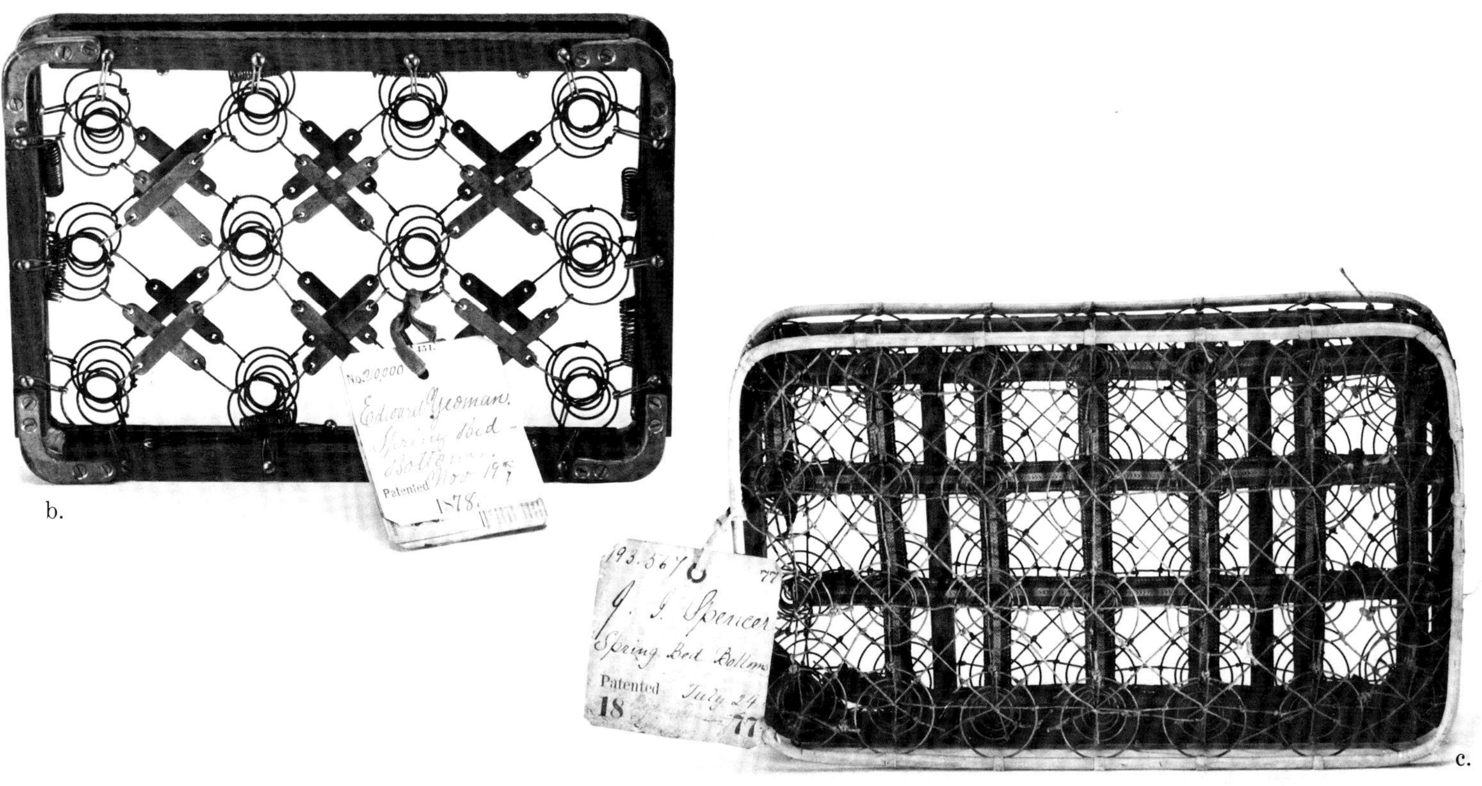

b.

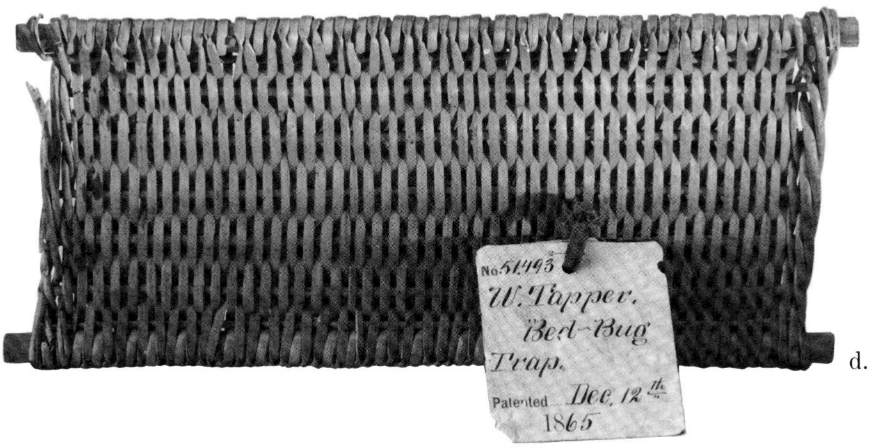

b.
Spring Bed Bottom
Edward Yeoman
Chicago, Illinois
November 19, 1878
Patent No. 210,000
PC

c.
**Improvement in Spring Bed-
    Bottoms**
James I. Spencer
New York, New York
July 24, 1877
Patent No. 193,567
PC

d.
Bed-Bug Trap
William Tapper
New York, New York
December 12, 1865
Patent No. 51,493
PC

a.
Portable Commode
William S. G. Baker
Baltimore County, Maryland
January 13, 1880
Patent No. 223,574
PC

b.
Improvement in Baby Walkers
John H. Headler
Brooklyn, New York
April 23, 1878
Patent No. 202,724
SI

c.
**Improvement in
  Enclosures or Playhouses for
  Children**
Hiram J. Parker
Rochester, New York
November 2, 1875
Patent No. 169,471
SI

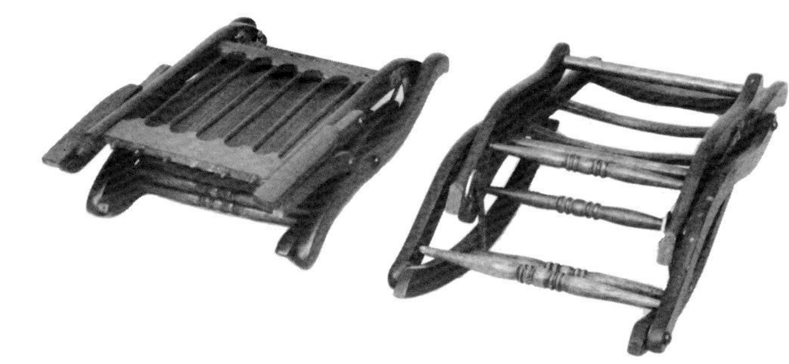

d.
Folding Chairs
Ephraim Tucker
Worcester, Massachusetts
May 25, 1880
Patent No. 228,144
PC, Photograph: PC

e.
Folding Rocking Chair
Ephraim Tucker
Worcester, Massachusetts
June 18, 1878
Patent No. 205,015
PC, Photograph: PC

a.

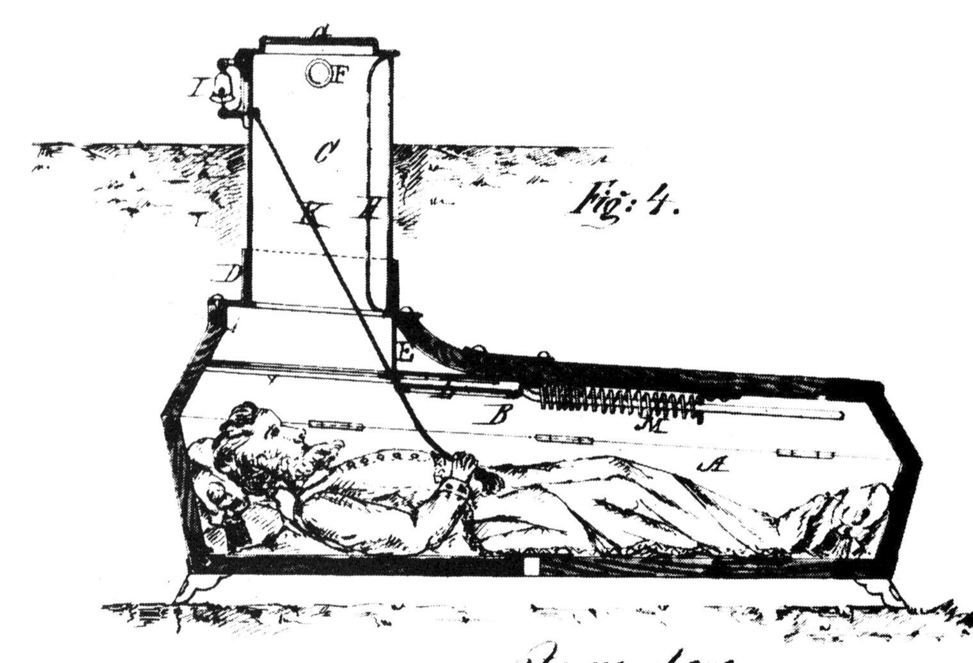

b.

a.

Improvement in Electric Baths
Mark W. House
Cleveland, Ohio
February 18, 1862
Patent No. 34,425
PC

b.

Improved Burial Case
Frank Vester
Newark, New Jersey
August 25, 1868
Patent No. 81,437
Courtesy National Archives

c.
Burial Urn
"The Home of a Deceased
   Friend
Died, February 22, 1848
Wallace
Peace to his Ashes"
[No number]
SI

d.
**Improvement in Copings and
   Coverings for Graves**
Charles H. Bassett
Derby, Connecticut
August 20, 1878
Patent No. 207,155
PC

d.

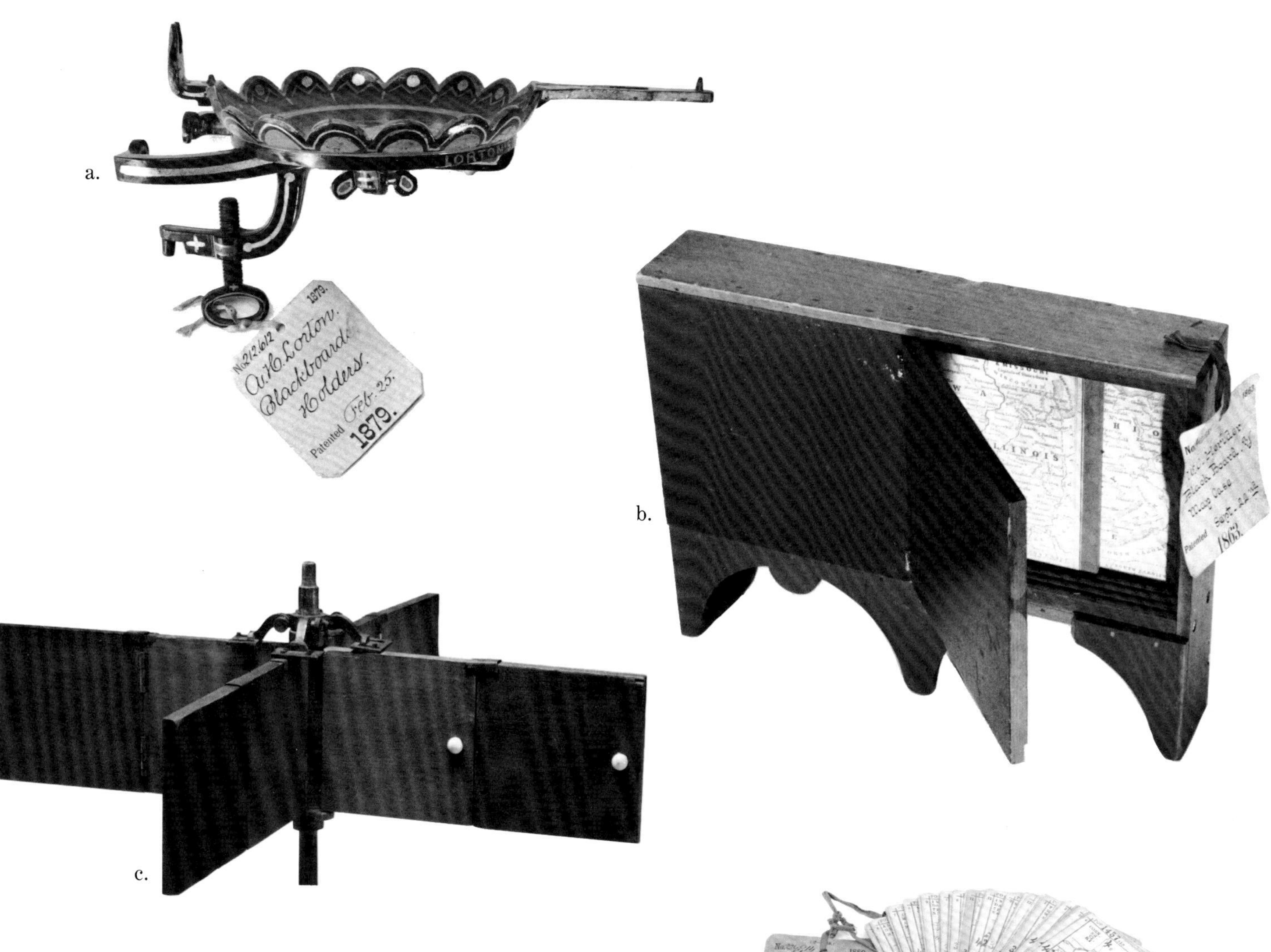

a.
Improvement in Blackboard-Holders
Alfred H. Lorton
New York, New York
February 25, 1879
Patent No. 212,612
PC

b.
Blackboard and Map Case
William C. Herider
Miami Town, Ohio
September 22, 1863
Patent No. 40,035
PC

c.
Improvement in Blackboards
Thomas J. Thorp
Buffalo, New York
January 21, 1873
Patent No. 135,019.
SI

d.
Geographical Playing Cards
Walter G. Read
San Francisco, California
July 13, 1880
Patent No. 229,914
PC, Photograph: Barry Korn

e.
School Desk
John Glendenning
Norwich, England
February 3, 1880
Patent No. 224,174
SI

f.
Improvement in School Desks
Charles H. Presbrey
Sterling, Illinois
January 21, 1873
Patent No. 135,154
SI

a.
Improvement in Inkstands
Herbert L. Andrews
Chicago, Illinois
March 31, 1868
Patent No. 76,138
PC

b.
Improvement in Inkstands
Thomas S. Shenston
Brantford, Ontario
November 25, 1873
Patent No. 144,929
PC

c.
Hand Printing Press
James N. Phelps
New York, New York
November 2, 1858
Patent No. 21,980
SI

d.

d.
**Combined Office Directory and
Desk**
William Walter
Washington, D.C.
March 21, 1876
Patent No. 175,215
SI

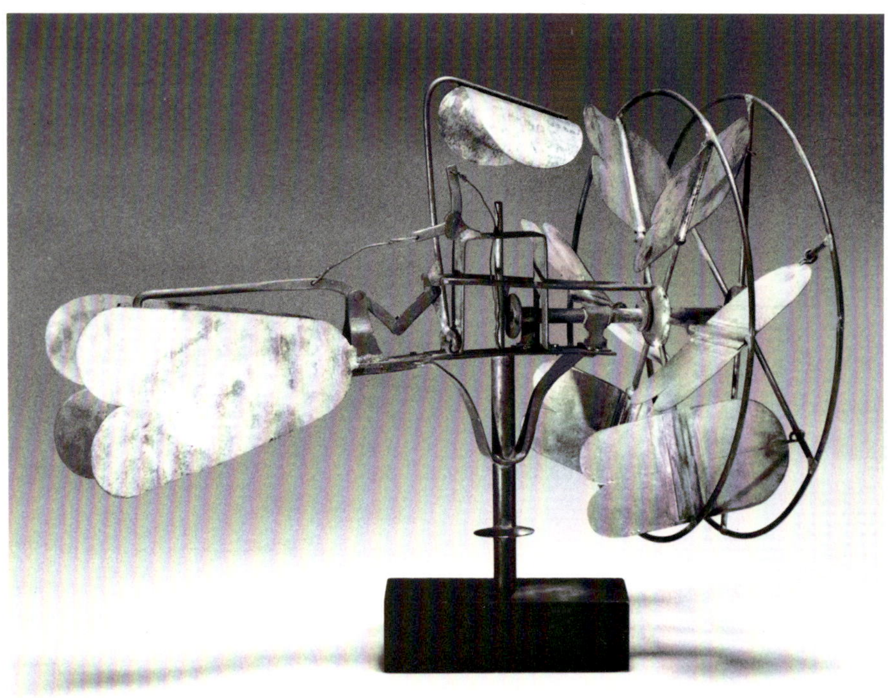

a.
Improvement in Windmills
Harvey M. Wood
Commerce, Michigan
December 2, 1879
Patent No. 222,340
SI

b.
**Improvement in Dog-Powered
  Treadmill**
Frederick K. Traxler
Dansville, New York
April 23, 1878
Patent No. 202,679
SI

c.

c.
**Improvement in Motors for Sewing
   Machines**
John Haworth
Philadelphia, Pennsylvania
April 30, 1878
Patent No. 203,035
SI

Power Production

Although the patent-model era coincided with the triumph of steam power in America, in both transportation and industry, there is ample evidence of a continuing devotion to "natural" forms of power (wind and water), and even to "low tech" oddities such as F. K. Traxler's treadmill. While the type of power required by the latter invention was unusual, the design—an endless track fastened to a flexible belt carried over two rollers, one of which was pivoted to permit any desired angle of inclination—has remained popular, and will of course be familiar to anyone who has used a stress-testing apparatus. The Wood windmill represents a conventional approach; the Haworth "water motor" is more novel in that its purpose was domestic—to operate a sewing machine. What is noteworthy about all three models shown here is that the discrete patented features were set in the context of the entire apparatus, in each case sufficiently simple to allow this. For the most part that was not true with improvements to steam-engine technology; patent models were ordinarily representative of a specific component or auxiliary in isolation.

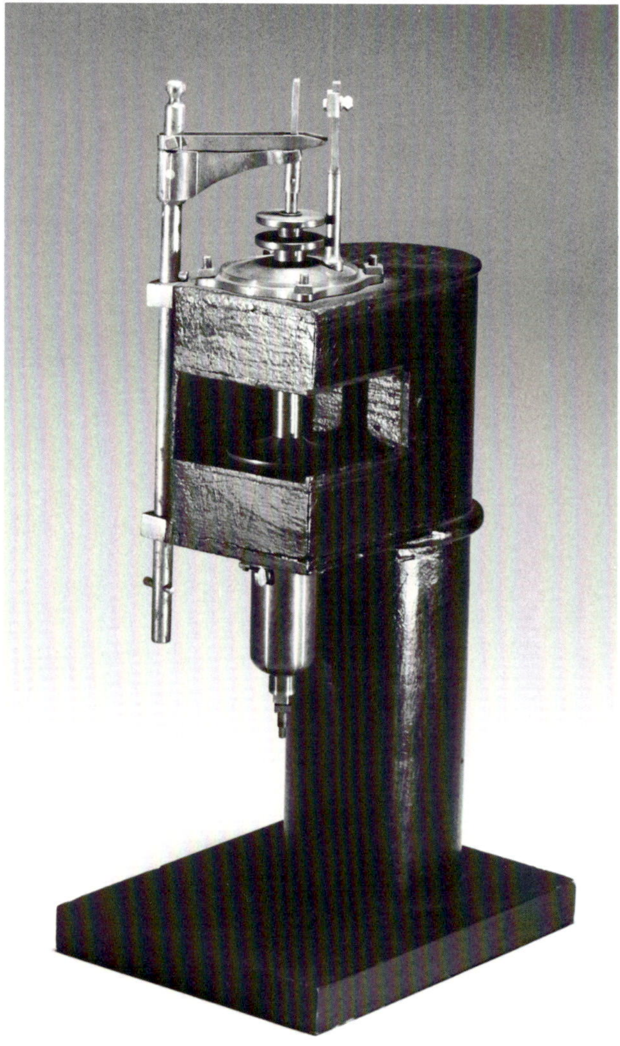

a.
Valve Gear for Steam Engines
George H. Corliss
Providence, Rhode Island
May 9, 1876
Patent No. 177,059
SI

b.
**Improvement in Lifting, Tripping,
  and Regulating the Closing of
  Steam Valves**
Frederick E. Sickels
New York, New York
May 20, 1842
Patent No. 2,631
SI

c.

Steam

While George Corliss, John Ericsson, and Frederick E. Sickels were all centrally involved in the improvement of power technology in the mid-nineteenth century, the name Corliss stood apart as (in Robert Vogel's words) "virtually synonymous with stationary steam power." The most significant of Corliss's many improvements was his "automatic drop-cutoff" for controlling the admission of steam to the cylinder, first patented in 1849 and refined many times subsequently. The model pictured here was submitted with an application for two patents, both of which were issued in 1876 the day before the opening of the Philadelphia Centennial, the exposition at which the great Corliss engine established its place in the history of American iconography.

In the realm of steam-engine technology, Sickels had patented his first improvement several years before Corliss. Although similar to Corliss's drop-cutoff, the device was not automatic and thus was more suitable for marine applications than for driving (factory) engines. Sickels tended to concentrate in the maritime realm (see also page 117), while Ericsson was an extremely versatile inventor who was involved not only in power but also in propulsion systems (the screw), in the design of ironclad warships, and in heavy ordnance (see page 110). He devoted a great deal of effort to the "caloric" or hot-air engine, the most spectacular application being in a 250-foot wooden sidewheeler, the *Ericsson*. Though it was a failure, it should be viewed, as Eugene Ferguson suggests, in the context of a series of "daring and audacious" innovations that characterized the 1850s, including the London and New York Crystal Palaces, John Roebling's Niagara Bridge, the *Great Eastern*, the Atlantic Cable, the Pennsylvania Railroad's Horseshoe Curve, and the five Collins Line transatlantic sidewheelers. Whatever harm Ericsson's reputation suffered as a result of the *Ericsson* was certainly recouped with the *Monitor* and, overall, he retains an image as one of the most admirable of American inventors. Even his patent models seem special; Ben Lawless notes that inventors such as Ericsson and Corliss produced "patent models that are skillfully made of first-rate materials, and sit among the other models with a kind of nonchalance as if to proclaim the superiority of the men whose names are affixed."

c.
Hot Air Engine
John Ericsson
New York, New York
December 20, 1845
Patent No. 4,317
SI

Rotary Steam Engines

a.
Rotary Steam Engine
Matthias Gabriel
Newark, New Jersey
August 6, 1867
Patent No. 67,527
SI

b.
Rotary Engine
Frederick F. Schofield
Oscoda, Michigan
September 19, 1876
Patent No. 182,291
SI

c.
Rotary Steam Engine
Charles Miller
Belleville, Illinois
May 3, 1859
Patent No. 23,852
SI

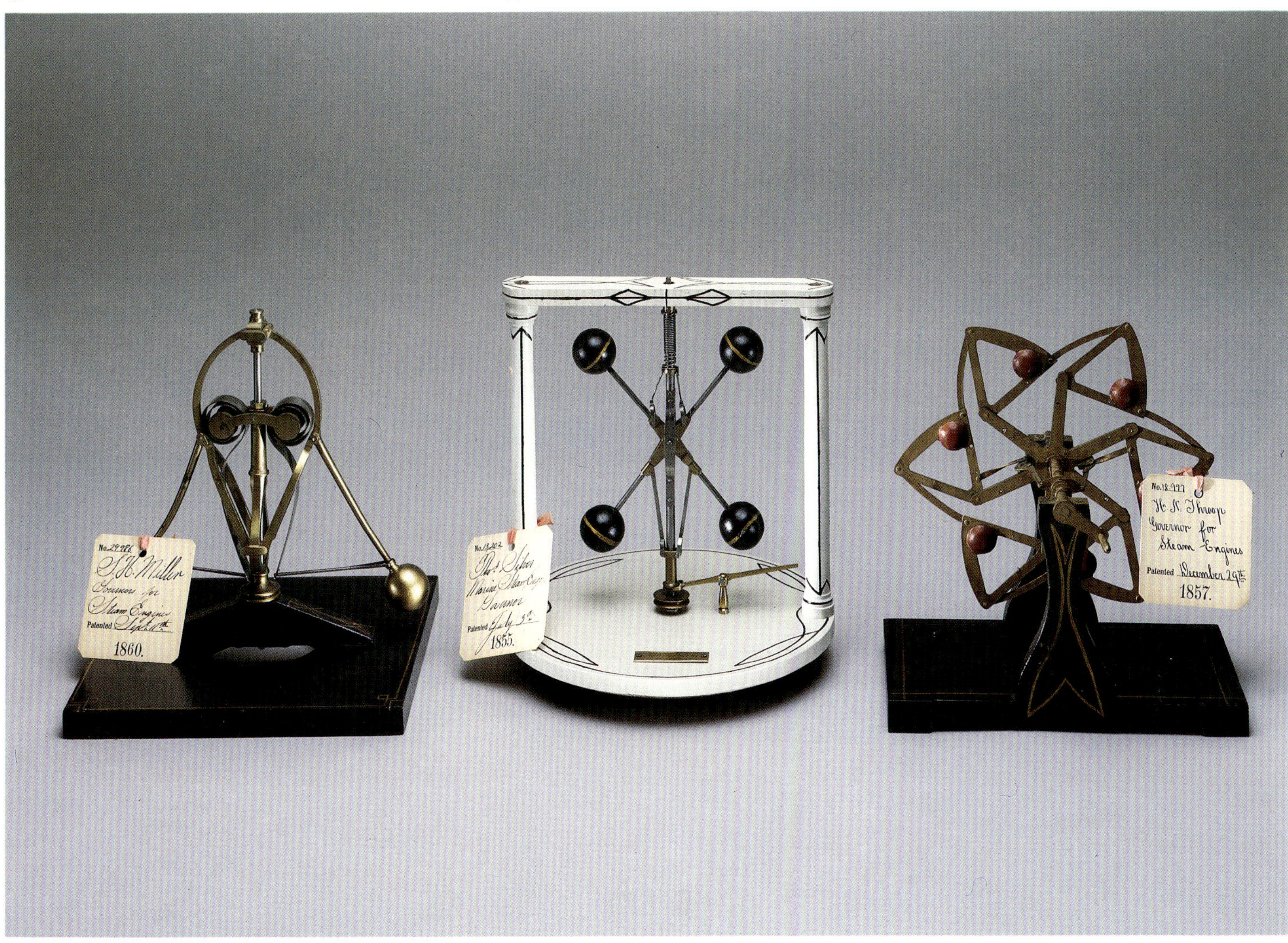

Governors

Essential to the operation of a stationary or marine steam engine is the governor, which maintains the speed constant even as external conditions such as load or steam pressure change. Governors are of a type of device now subsumed under the science of cybernetics, whose critical concept, "feedback," was described by Norbert Weiner as a capability "to adjust future conduct by past performance." Feedback devices are ubiquitous in our everyday lives; common examples include the thermostat, the float valve, and—most closely akin to a steam-engine governor—automobile "cruise control."

The models pictured here all stem from James Watt's discovery that centrifugal force swings a pair of rotating flyballs outward as speed increases, and that mechanical links between the rods suspending the flyballs and the steam inlet valve will admit more steam as engine speed slows, less as it accelerates. Each of these examples was designed for marine engines, which, because they are subject to rolling and pitching, require governors operable in orientations other than strictly vertical. Of special note is Throop's, which embodies a remarkably subtle feature: Not only does centrifugal force displace the flyballs radially outward as engine speed increases, they also move tangentially under acceleration or deceleration—action thereby responding both to speed and to its rate of change, a characteristic known in modern control engineering as "proportional plus derivative response." Silver's governor, though perhaps less elegantly conceived, brought him considerable commercial success, with his customers ultimately including the navies of both France and Britain.

d.
Improvement in Governors for Steam-Engines
S. H. Miller
Hanoverton, Ohio
September 11, 1860
Patent No. 29,986
SI

e.
Governor for Marine Steam Engines
Thomas Silver
Philadelphia, Pennsylvania
July 3, 1855
Patent No. 13,202
SI

f.
Governor for Steam Engines
H. N. Throop
Pultneyville, New York
December 29, 1857
Patent No. 18,997
SI

a.

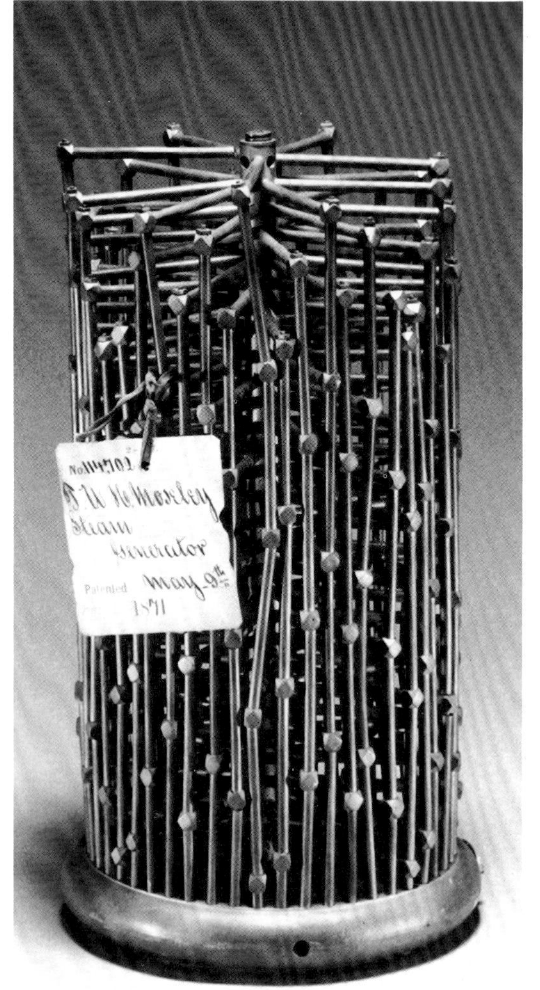

b.

c.

Engine Components & Steam-Powered Auxiliaries: Style and Aesthetics

These models suggest something of the stylistic range of what was in effect the leading edge of American technology in the 1860s and 1870s. The Worthington pump is a relief panel of the sort Kendall Dood calls a "plaque," a facsimile of mechanical principles in a cutaway mode and in wood rather than metal. The Sturtevant blower, on the other hand, is perfectly representational, simply a reduced-scale version of the actual device. In the case of certain accessories such as injectors, the model might in fact *be* identical to the production device, a situation which held true in other realms

such as hand tools, timekeeping devices, surgical and dental instruments, and various kinds of latches and fasteners. The Westinghouse pump is slightly abstract, lacking some details that might have been included, and the Moseley boiler has a wild, almost surrealistic aspect. The condensers are miniatures devoid of context.

a.

Condenser
George K. Gluyas and
   Washington R. Pitts
San Francisco, California
October 1, 1872
Patent No. 131,779
SI

b.

Steam Generator
Thomas W. H. Moseley
Hyde Park, Massachusetts
May 9, 1871
Patent No. 114,702
SI

c.

Steam-Engine Condenser
Francis B. Stevens
New York, New York
November 3, 1863
Patent No. 40,510
SI

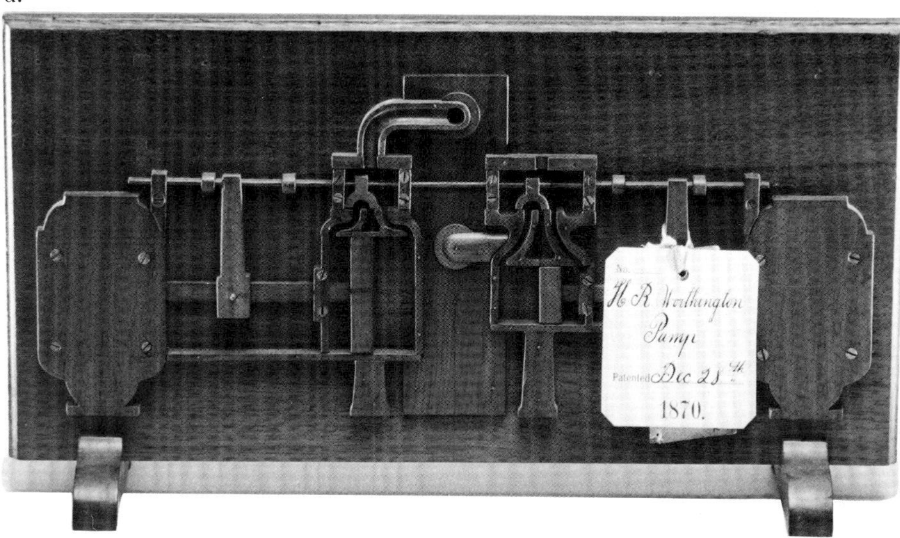

d.

Duplex Steam Pump
Henry R. Worthington
New York, New York
June 20, 1871
Patent No. 116,131
SI

e.

**Improvement in Compound Air-
Heaters and Steam-Condensers**
B. F. Sturtevant
Jamaica Plain, Massachusetts
February 22, 1870
Patent No. 100,242
SI

f.

Steam Engine and Pump
George Westinghouse, Jr.
Pittsburgh, Pennsylvania
August 30, 1870
Patent No. 106,899
SI

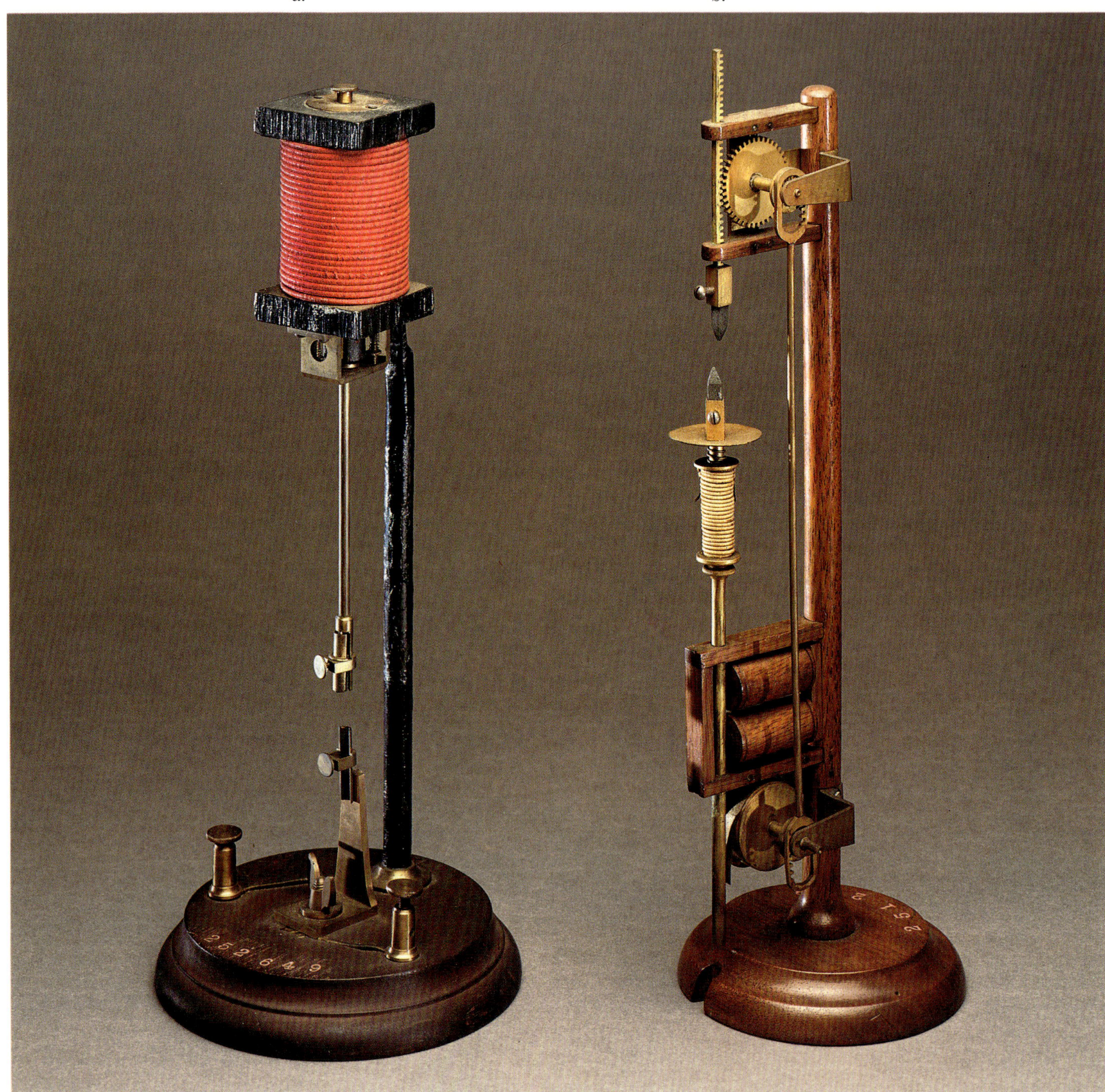

a.
Improvement in Electric Lamps
Charles F. Brush
Cleveland, Ohio
May 7, 1878
Patent No. 203,411
SI

b.
Improvement in Electric Lights
Matthias Day, Jr.
Mansfield, Ohio
February 24, 1874
Patent No. 147,827
SI

c.

d.

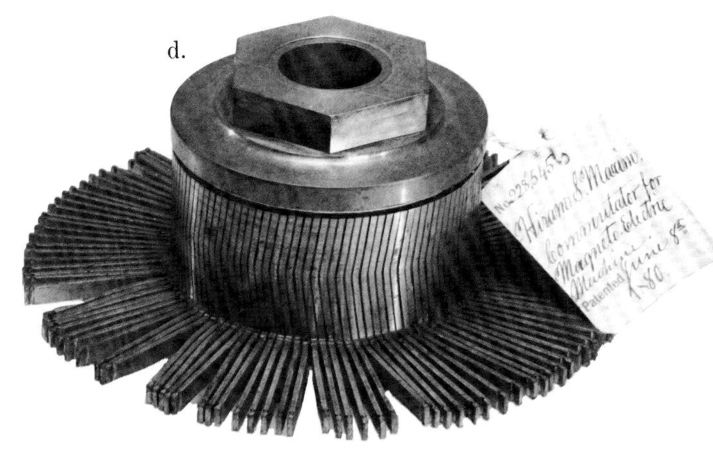

f.

e.

e.
Improvement in Magneto Electric Machine
Charles F. Brush
Cleveland, Ohio
April 24, 1877
Patent No. 189,997
SI

c.
Dynamo-Electric Machine
Hiram Maxim
New York, New York
June 8, 1880
Patent No. 228,544
SI

d.
Commutator for Magneto Electric Machines
Hiram Maxim
New York, New York
June 8, 1880
Patent No. 228,545
SI

f.
Carbonizer
Thomas A. Edison
Menlo Park, New Jersey
October 18, 1881
Patent No. 248,423
PC, Photograph: PC

Electricity

With one exception, all the technologies derived from electrical science were barely developed at the time the model requirement was dropped from the rules of the Patent Office. The inventive genius of the 1870s was symbolized in the "hoarded power" of the Corliss Centennial Engine; the few electrical exhibits in 1876 attracted little attention. It was at the next great American world's fair, the Columbian Exposition in Chicago in 1893, that electrical technology commanded center stage just as steam and factory machinery had done at Philadelphia.

The one electrical technology to have matured early was telegraphy. Samuel F. B. Morse had inaugurated his commercial system in 1844, and within

less than a decade his company alone was operating some 13,000 miles of line. By the 1870s all the states of the Union were linked together, and, via submarine cable, New York was connected to London and far beyond. Telegraph innovation had scarcely been a Morse monopoly; indeed, all of the inventors whose names appear with the models on pages 74 and 75 had played a significant role, especially Phelps with his printer. Elisha Gray's special concern had been a system of tuned transmitters and receivers by means of which he hoped to be able to send multiple messages simultaneously over the same wire. Gray's work along these lines had suggested to him the possibility of transmitting voices telegraphically, but it was a rival inventor, Alexander Graham Bell—likewise interested in harmonic telegraphy—who ultimately patented the first telephone.

While Bell, Gray, and other telephone pioneers such as Emile Berliner were doing business with the Patent Office early enough to have submitted a number of models, and the first telephone exchange (in New Haven) opened in January 1878, telephony remained at a very early stage in 1880.

Electric power was also at an early stage of development. Electrical communication required so little current that chemical batteries sufficed. Both magnetos (permanent-magnet generators) and battery-powered electric motors had been developed as early as the

a.
Self-Winding Telegraph Registers
James J. Clark
Philadelphia, Pennsylvania
October 18, 1853
Patent No. 10,128
SI

b.
Improvement in Electric Telegraph
Samuel F. B. Morse
Poughkeepsie, New York
May 1, 1849
Patent No. 6,420
SI

c.
Improvement in Indicating Telegraph
Lucius G. Curtiss
Cincinnati, Ohio
January 16, 1849
Patent No. 6,040
SI

1830s, but in the 1870s there were only faint indications that electricity might eventually challenge steam as a major source of power. The dynamo, a device that converts other forms of power to electricity, proved its utility with arc lighting, which required considerable current. However, its full potential only came to be realized when Thomas Edison, already experimenting with incandescent light, turned his attention to the idea of the "central station." Here a power source such as steam could be linked to a dynamo, and the electricity generated could then be sent through wires for use at diverse points beyond.

While apparatus such as dynamos have a mechanical component—generators do, after all, need to be driven by engines or turbines of some sort—electrical technology is fundamentally intangible. Such inventions as the Bell telephone apparatus, the Maxim models, even the Edison carbonizer, have nothing mechanically obvious about them. Had the time not come to terminate the model requirement for simple administrative reasons, it would soon have begun to fade away for purely technical reasons—that is, the impossibility of modeling intangibles.

d.
Electra Harmonic Telegraph
Elisha Gray
Chicago, Illinois
February 15, 1876
Patent No. 173,618
SI

e.
Magneto-Electric Speaking Telephone
Ansen Phelps
Brooklyn, New York
August 19, 1879
Patent No. 218,684
SI

f.
Improvement in Telegraphy
Alexander G. Bell
Salem, Massachusetts
March 7, 1876
Patent No. 174,465
SI

g.
Contact Telephone
Emile Berliner
Boston, Massachusetts
May 24, 1881
Patent No. 241,912
SI

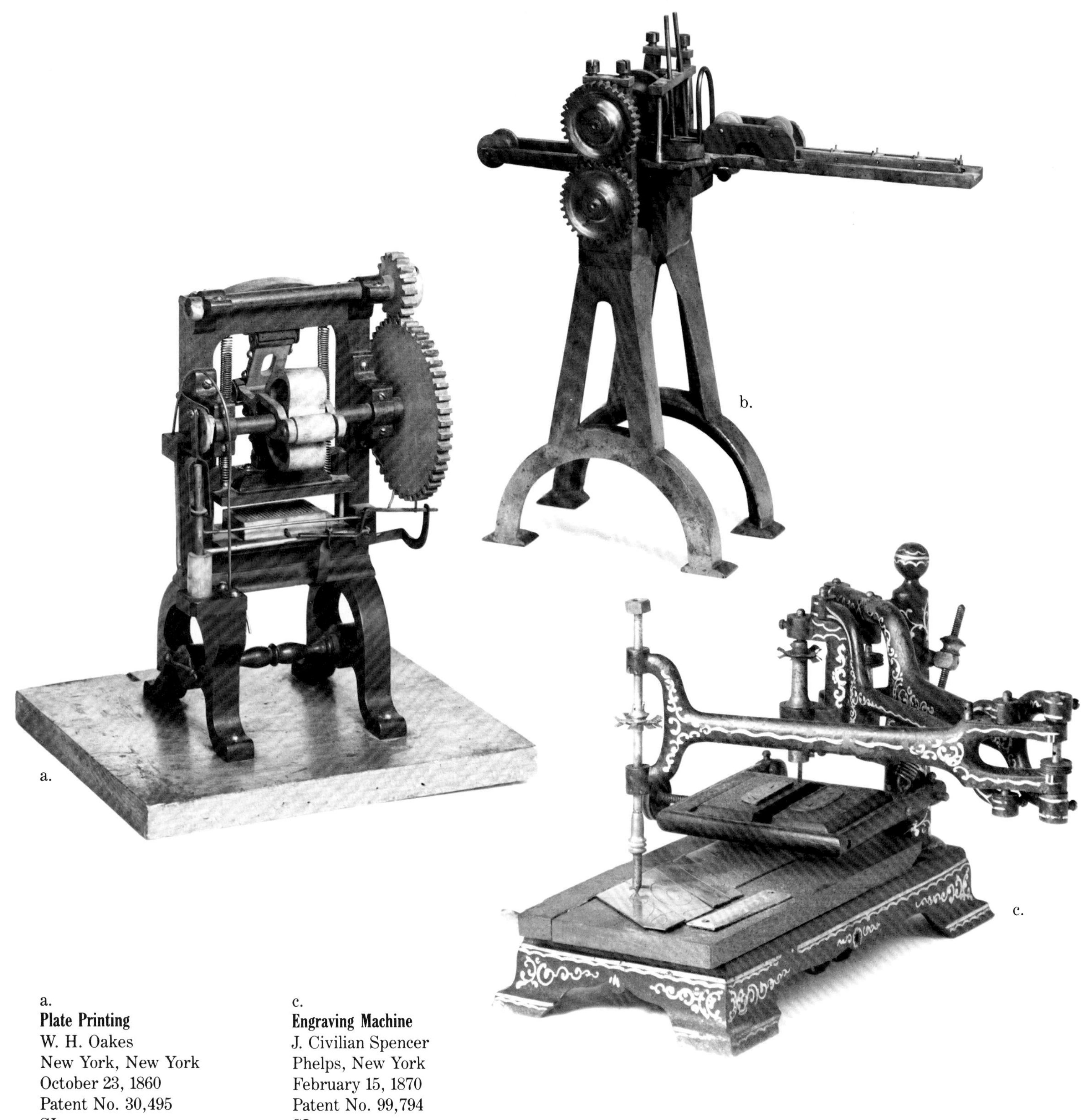

a.
Plate Printing
W. H. Oakes
New York, New York
October 23, 1860
Patent No. 30,495
SI

b.
**Printing Press for Addressing
  Newspapers**
George Henderson
Allegheny, Pennsylvania
September 6, 1859
Patent No. 25,363
SI

c.
Engraving Machine
J. Civilian Spencer
Phelps, New York
February 15, 1870
Patent No. 99,794
SI

d.

Visual Communication

Besides the models for Bell's "telephonic telegraph" and the variety of earlier telegraphic technologies, the patent models range through a vast array of inventions for copying, recording, and reproducing information: for setting and printing type, for exposing and printing photographic plates, and for mechanizing the process of writing.

As would seem to befit their leading role in communications technology, patentees of improvements to printing presses submitted models that are among the most professionally crafted of all; they are almost without exception elegantly detailed miniatures on the order of the Wilkinson apparatus of 1859.

d.
Printing Press
Jeptha Avery Wilkinson
Brooklyn, New York
August 9, 1859
Patent No. 25,069
SI

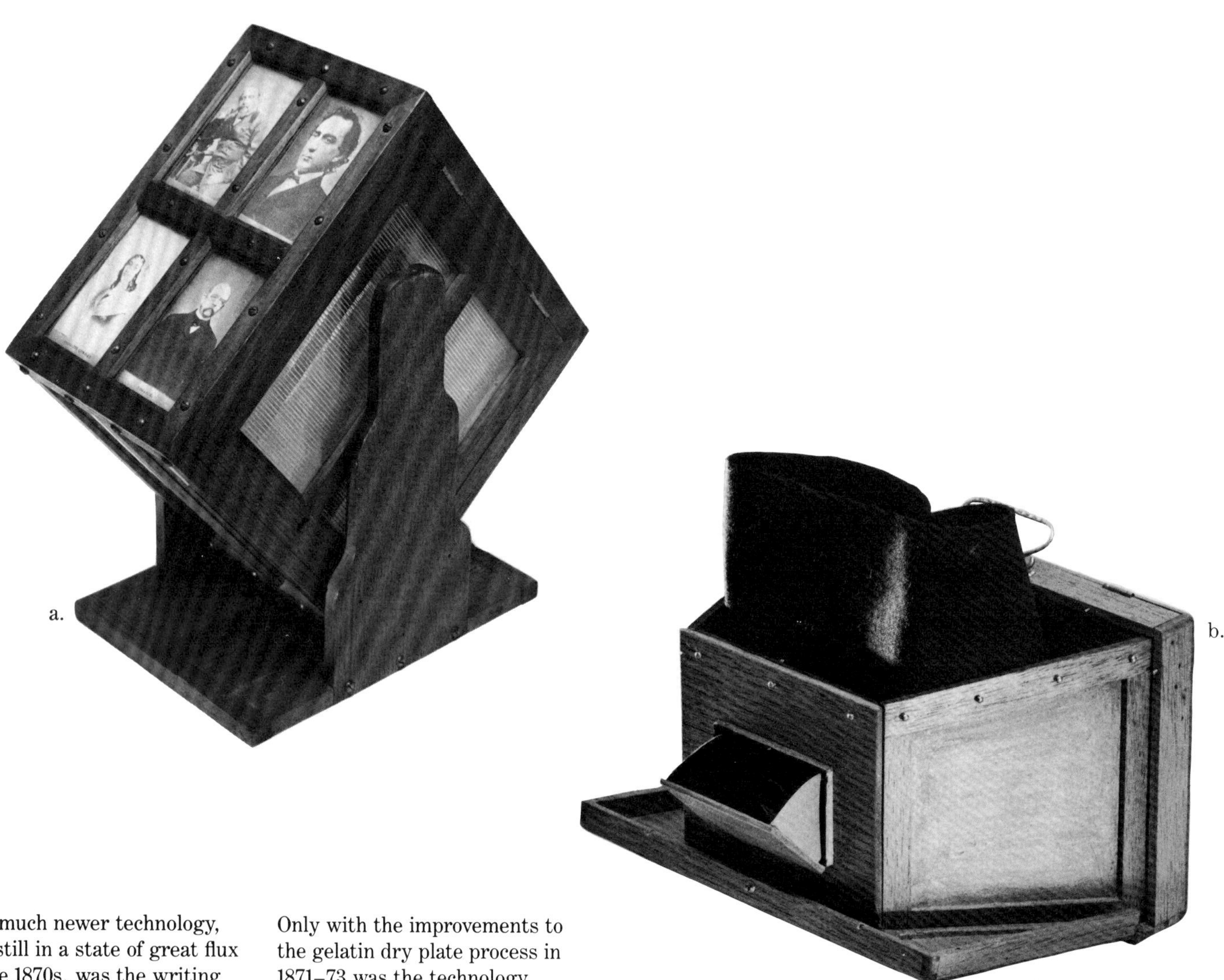

A much newer technology, and still in a state of great flux in the 1870s, was the writing machine. Although American precursors may be found as early as 1843 (a patented "mechanical chirographer"), and Alfred E. Beach, co-proprietor of *Scientific American*, built several experimental models around mid-century, a truly practical typewriter was not invented until after the Civil War. Typewriters were being sold commercially by 1874, but even when the Hall and Washburn patents were issued there was as yet no assured market for such devices, nor—as is obvious from these models— had any design conventions been established.

With photography the situation was somewhat different, for this was a well-developed professional craft by the 1870s.

Only with the improvements to the gelatin dry plate process in 1871–73 was the technology made available to the amateur who lacked expertise in chemistry. Today, the technology is so accessible that Americans snap some eight million shots every day. Although advances in the mechanical aspects of photographic equipment have been important, the most critical improvements in photography have been in the realm of chemical processes, which cannot be shown in the form of a model. A miniature printing press, whatever the discrete patented features, is "something entirely and immediately apprehendable to the senses"; but how could anyone "model" what happened when a sensitized plate was exposed to light?

a.
Improvement in Photographic Picture-Holder
Alfred Thomas
Whitpain, Pennsylvania
December 22, 1874
Patent No. 157,945
SI

b.
Improvement in Portable Cameras
C. Alfred Agren
New York, New York
March 3, 1874
Patent No. 148,019
SI

c.

c.
Type Writing Machine
William A. Hall
St. Louis, Missouri
July 30, 1878
Patent No. 206,439
DC, Photograph: Bill Jacobson

d.
Type Writing Machine
Charles A. Washburn
San Francisco, California
November 8, 1870
Patent No. 109,161
DC, Photograph: Bill Jacobson

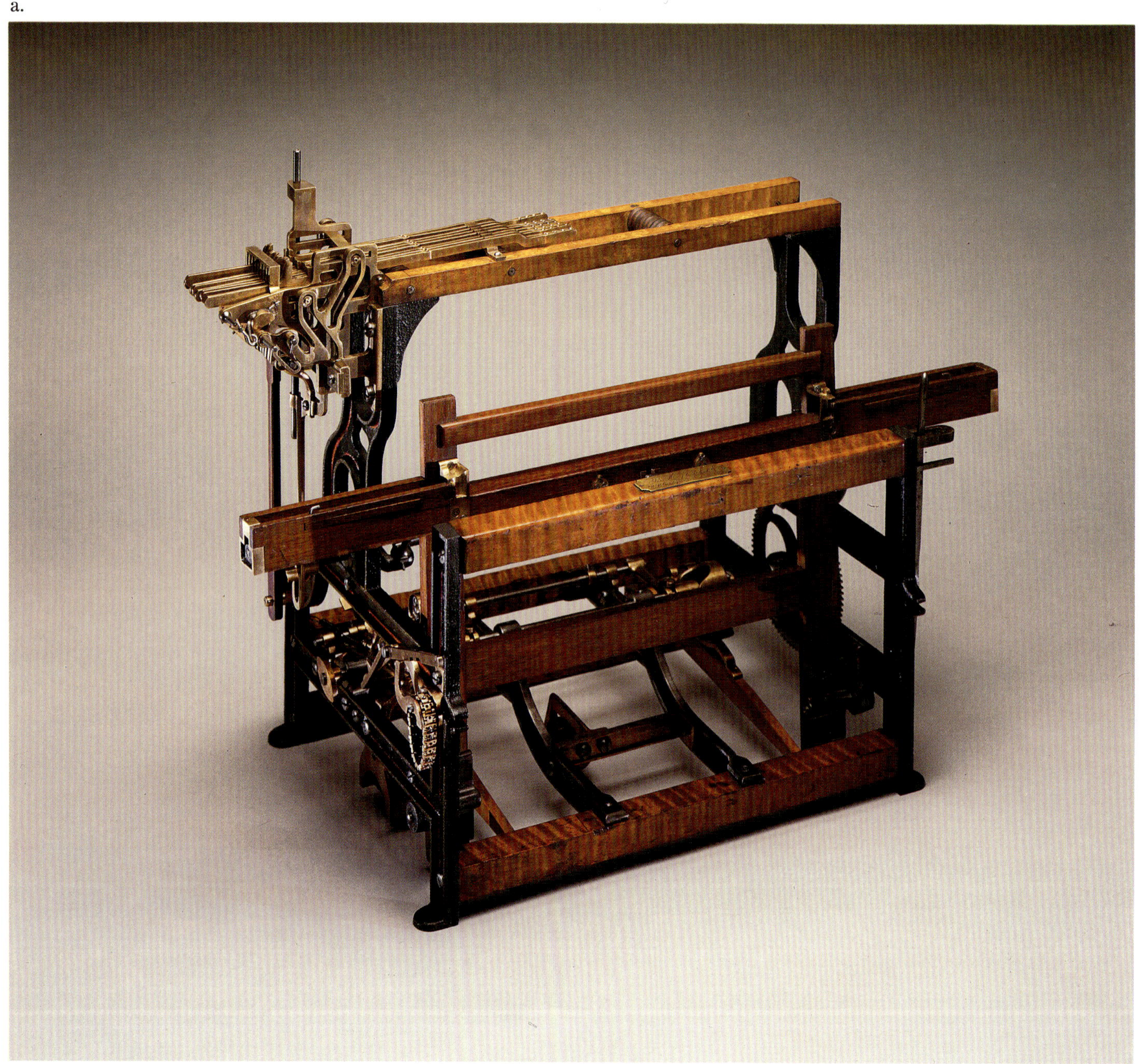

a.
**Improvement in Heddle Motion for
  Looms**
Improvement in Looms
Barton H. Jenks
Robert B. Goodyear
Philadelphia, Pennsylvania
May 22, 1866
Patent Nos. 55,010, 55,011
SI

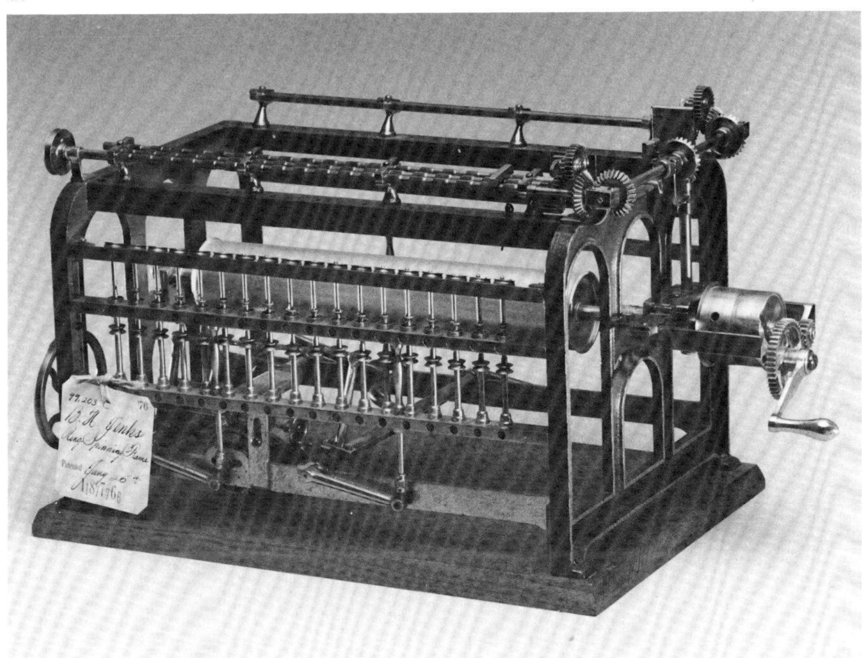

Textiles and Apparel

Textile manufacture was the first in America to be mechanized, and, although the industry had its ups and downs as a result of economic depressions and wartime disruptions, it persisted as a triumph of industrial capitalism. Steve Dunwell has written that "New England textiles enjoyed a century of unchallenged supremacy as a result of having monopolized the nation's motive power, venture capital, and mechanical skill." In the realm of mechanical skill, a second and third generation of inventors continued improving the machinery (improving it from the mill-owners' perspective, that is), and patent models as a rule were finely detailed miniatures with the same standard of craftsmanship as printing presses.

Though Jenks and Goodyear (not Charles, the "martyr to rubber") were significant inventors in their own realm, the most famous name on this page is Erastus Bigelow, one of the *Men of Progress*. Bigelow had patented a power loom for coach lace as early as 1837, but was chiefly famous for adapting the principles of the power loom to the production of pile fabrics, wire cloth, and Brussels, Wilton, and velvet carpeting. "A name on the door rates a Bigelow on the floor," ran the slogan. The patent model shown here, which accompanied one of Bigelow's later patent applications, dates from 1876, three years before his death.

b.
Loom
Erastus B. Bigelow
Boston, Massachusetts
May 30, 1876
Patent No. 177,920
SI

c.
Machinery for Manufacturing Felt Cloth without Spinning and Weaving
Thomas R. Williams
London, England
December 14, 1840
Patent No. 1,897
SI

d.
Ring Spinning Frame
Barton H. Jenks
Bridesburg, Pennsylvania
January 25, 1870
Patent No. 99,203
SI

Among the array of satellite industries to textile production were fulling (a process both chemical and mechanical) coating with "plastic compositions" (making, in the phraseology of Henry Loewenberg's patent, "imitation articles of leather or straw, enameled cloth, trimmings, embroideries, &c."); and hatmaking (readily mechanized—there were more than 120 pertinent patents by mid-century). Shoemaking, on the other hand, was still largely hand-work at mid-century. Though production was considerable—Lynn, Massachusetts, alone accounting for 2.5 million pairs annually—the shoe industry required a national work-force of 150,000 to 200,000. Not until 1858 was a machine patented capable of sewing soles to uppers, and even in the 1870s the level of mechanization still remained low.

a.
**Improvement in Apparatus for
Coating Fabrics with Plastic
Compositions**
Henry Loewenberg
Charlottenburg, Prussia
April 11, 1876
Patent No. 176,019
PC

b.
**Improvement in Chemical Processes
    for Fulling Vegetable and Other
    Textures**
John Mercer
Oakenshaw, England
August 19, 1851
Patent No. 8,303
PC

c.
Improvement in Fulling-Mills
Joseph H. Trainor
Springfield, Vermont
May 5, 1874
Patent No. 150,663
PC

a.

b.

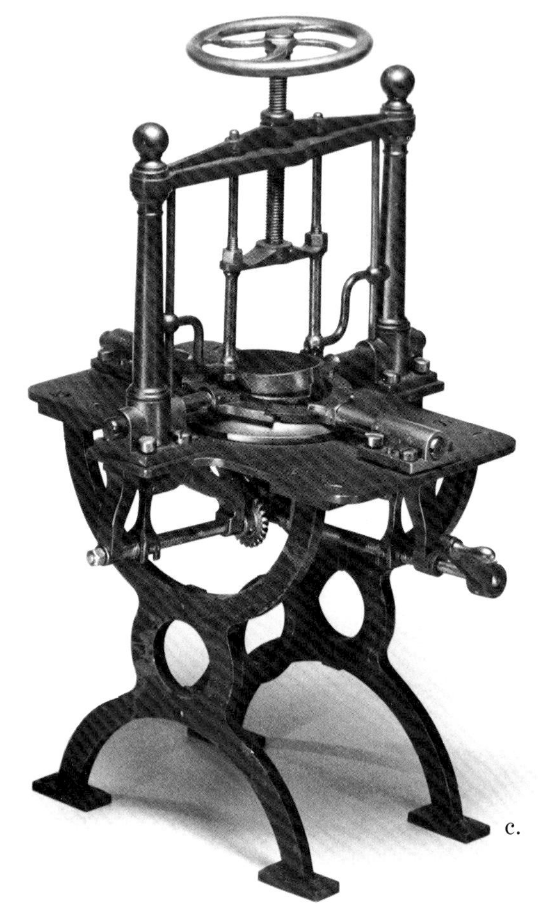

c.

a.
Hat-Block
William W. Cumberland
Newark, New Jersey
February 19, 1861
Patent No. 31,442
PC

b.
Machine for Hardening Hat Bodies
Seth Boyden
Newark, New Jersey
August 30, 1859
Patent No. 25,300
PC

c.
Hat Shaping Machine
J. P. Morlot
France
February 1, 1870
Patent No. 99,458
SI

d.
Machine for Lasting Boots and Shoes
Charles H. Trask
Lynn, Massachusetts
and Henry F. Wheeler
Boston, Massachusetts
June 23, 1873
Patent No. 142,657
SI

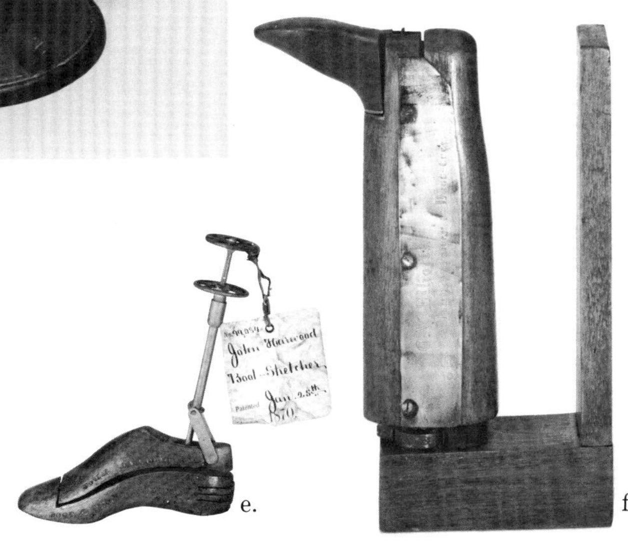

e.

f.

g.

e.
Improved Boot-Stretcher
John Harwood
Albany, New York
January 25, 1870
Patent No. 99,084
PC

f.
Improvement in Boot-Trees
Orson V. Elliott
Mansfield, Pennsylvania
July 1, 1873
Patent No. 140,490
PC

g.
Improvement in Nailing-Lasts for Boots and Shoes
Joseph A. Safford
Winchester, Massachusetts
September 24, 1872
Patent No. 131,565
PC

a.
**Improved Apparatus for Rendering
  Tallow**
John J. Eckel and Isaac S.
  Schuyler
New York, New York
August 14, 1866
Patent No. 57,104
PC

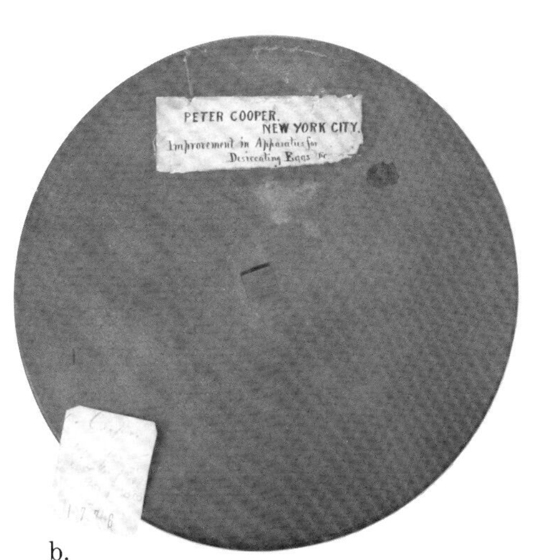

Chemical Manufactures

While the textile mill became emblematic of the plague of industrial hazards in the nineteenth century, other industries attained greater infamy in the annals of pollution. Long before anybody showed any concern for airborne particulates or contaminated water, inventors were attempting to subdue olfactory nuisances, especially those associated with rendering and the refining of petroleum. Joshua Merrill addressed the problem of the "persistent disagreeable smell" of lubricants distilled from crude oil. Eckel and Schuyler, who were from New York, where there had been attempts to crack down on rendering plants, sought to suppress "the dissemination of vapors and unpleasant odors" from rendering operations by injecting the vapors "into the fire-chambers under the kettle." Still, there were certain urban undertakings—gas generation, for example—that citizens simply had to learn to tolerate.

While three of the models pictured here have the sort of multitubular complexity that one associates with chemical works, Peter Cooper's "improvement in that class of apparatus for the desiccation of eggs, glue, and similar fluid or semi-fluid substances" is probably the least complex patent model in this book, consisting merely of a "non-metallic mineral plate or disk, formed with unglazed tenacious surfaces, adapted to collect and retain . . . the liquid film for desiccation." This is the only known model associated with a Cooper patent, and it suggests how unprepossessing a model of a significant invention could be.

b.
Improvement in Plates for Desiccating Eggs
Peter Cooper
New York, New York
January 25, 1876
Patent No. 172,611
PC

c.
Improved Apparatus for Carbureting Air
Warren A. Simonds
Boston, Massachusetts
March, 21, 1865
Patent No. 46,976
SI

d.
Improved Process for Deodorizing Heavy Hydrocarbon-Oils
Joshua Merrill
Boston, Massachusetts
May 18, 1869
Patent No. 90,284
PC

a.

b.

Industrial Detergents

Morse's apparatus for "separating dust" and Crum's for "renovating" feathers had similar purposes. The former, which was patented in at least seven foreign countries as well as the U.S., was intended for flour mills and other factory operations whose efficiency was hampered by the copious discharge of airborne particles. The latter was designed to precipitate and flush away dirt while feathers were being steamed preparatory to the making of such products as hats and dusters.

a.
Dust-Collector
Orville M. Morse
Jackson, Michigan
May 14, 1889
Patent No. 403,362
PC

b.
Improvement in Feather Renovators
George H. Crum
Ladiesburg, Maryland
October 8, 1878
Patent No. 208,673
PC

c.

d.

Processing Pharmaceuticals

The models that accompanied applications for patents on various means of mechanizing drug production tended to have a spic-and-span aspect, which was no doubt calculated. "Manufacturing chemists" presumably ran the cleanest of factory operations, just as the operations to which the models on the opposite page pertained were exceedingly dirty.

c.
Apparatus for Sugar-coating Confectionery, Pills, etc.
William Cairns
Jersey City, New Jersey
February 16, 1875
Patent No. 159,899
SI

d.
Machines for Putting Up Seidlitz Powders
Charles R. Doane
Brooklyn, New York
May 8, 1877
Patent No. 190,564
SI

a.
**Improvement in Machines for
    Making Candy**
Thomas and George M. Mills
Philadelphia, Pennsylvania
February 14, 1871
Patent 111,765
SI

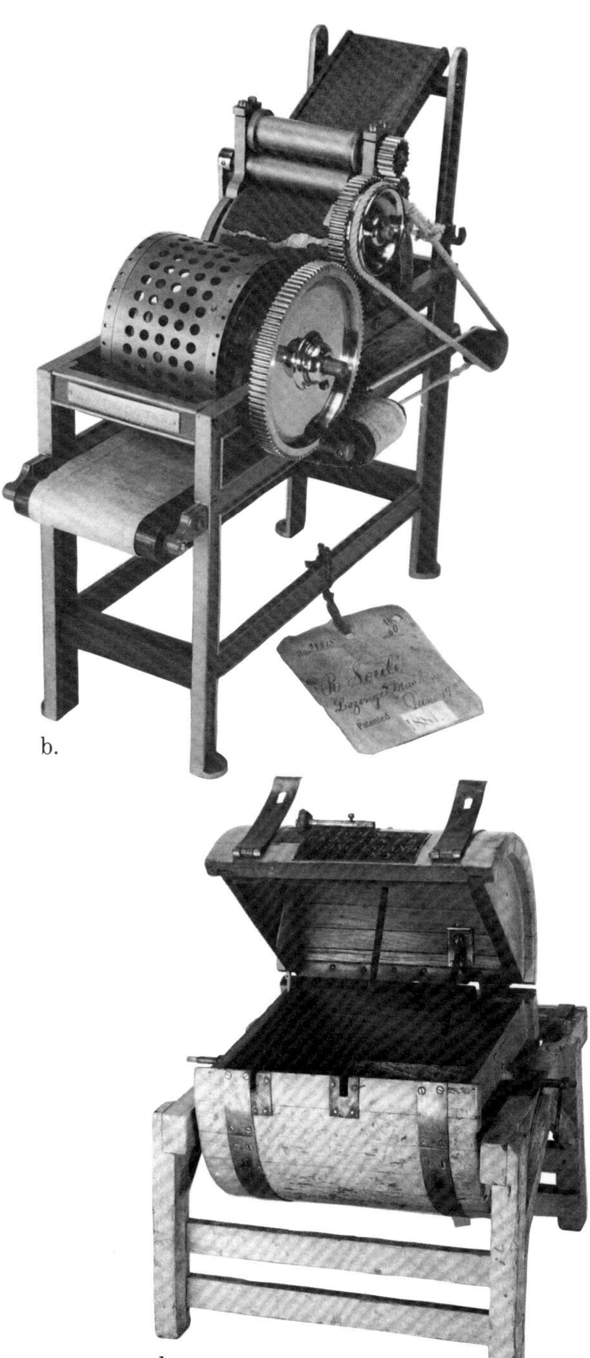

b.

d.

c.

Processing Food

The Sowle and Mills models pictured here are functionally quite similar, though differently scaled. Thomas Mills & Bro., which held the patent on the hand-operated machine, was a large-scale confectioner in Philadelphia; the Sowle patent was actually in the name of Rhoda Sowle, "administratrix of the estate of David Sowle, deceased" (a legal situation not uncommon with respect to patents). Henry John Heinz, whose inventions ultimately included one of the most memorable of advertising slogans—"57 Varieties"—patented this vegetable sorter three years after the founding of F. & J. Heinz, which became the H. J. Heinz Co. in 1888.

b.
Improvement in Lozenge-Machines
Rhoda Sowle
Fall River, Massachusetts
June 19, 1860
Patent No. 28,815
PC

c.
Vegetable Assorter
Henry John Heinz
Sharpsburg, Pennsylvania
February 4, 1879
Patent No. 212,000
SI

d.
Dough Mixer
Thomas Holmes
Brooklyn, New York
June 15, 1869
Patent No. 91,335
DF

a. b.

Machine Tools

Woodworking and metalworking machines are both central to the advent of the American system of manufacture, which is one of the most significant topics in the history of the nineteenth century, and one that defies a brief summary. Suffice it to say here that American machine tools as a whole outshone even the Corliss Engine at the Centennial, and by the 1890s the popular recognition of their role in America's rise to world power was so universal that the Patent Office actually built replicas of the Blanchard lathe model and the Rinewalt planer (the originals having been destroyed by fire) to send to the Columbian Exposition along with a display of genuine patent models.

While models for woodworking machines tended to be a bit rough and ready, those for precision metalworking machines—"machines to make machines"—were usually perfectly detailed working miniatures. The two pictured on the facing page accompanied patent applications that were accepted shortly before the Centennial, when the world first began to comprehend the industrial potential of the U.S. and its technological basis. Edwin Battison has written that "what became perfectly clear at Philadelphia was that the stunning displays of American mass-produced products were the offspring of similarly superb American wood- and metal-working machines."

The precision and complexity of such machines made models very costly even for experimental purposes, since most machine-tool designers preferred to work to full scale and from drawings—which were, after all, the key to presenting a successful patent application. Here, then, even as the mechanical era neared its apogee, one senses the waning relevance of patent models.

c.

d.

a.
Lathe for Turning Irregular Forms
Thomas Blanchard
Springfield, Massachusetts
Patent Issued January 20, 1820
Patent Office Replica, 1892
SI

b.
Wood-Planing Machine
George Rinewalt, Jr.
Pendleton, Indiana
July 30, 1861
Patent No. 32,926
Patent Office Replica, 1892
SI

c.
Improvement in Milling-Machines
William Krutzsch
Dayton, Ohio
August 17, 1875
Patent No. 166,704
SI

d.
Improvement in Gear-Cutting Machines
John A. Peer
San Francisco, California
July 21, 1874
Patent No. 153,370
SI

Matters of Identity

For diversified manufacturing processes beyond the "machine shop"—for machines designed not to make machines but to make containers and building components from bricks to nails to shingles—the typical model was a fully operational miniature. There is something intrinsically attractive about miniaturization, and one can perceive from the worn condition of such models that they were frequently handled and made to "work." One result of their popularity is the frequency of missing tags, so that, unless the patentee inscribed or lettered an identifying name or phrase directly on the model, one is confronted with an anonymous artifact.

This plight need not be permanent, for in time it should be possible to identify many such models (provided they are *patented* models) by means of a systematic comparison with the drawings. Some, however, are fated to remain anonymous and to be appreciated just for themselves.

a.

a.
Brick Molding Machine
[No number]
PC

b.
Nail-Machine
F. Davison
Richmond, Virginia
July 21, 1868
Patent No. 80,150
PC, Photograph: Barry Korn

b.

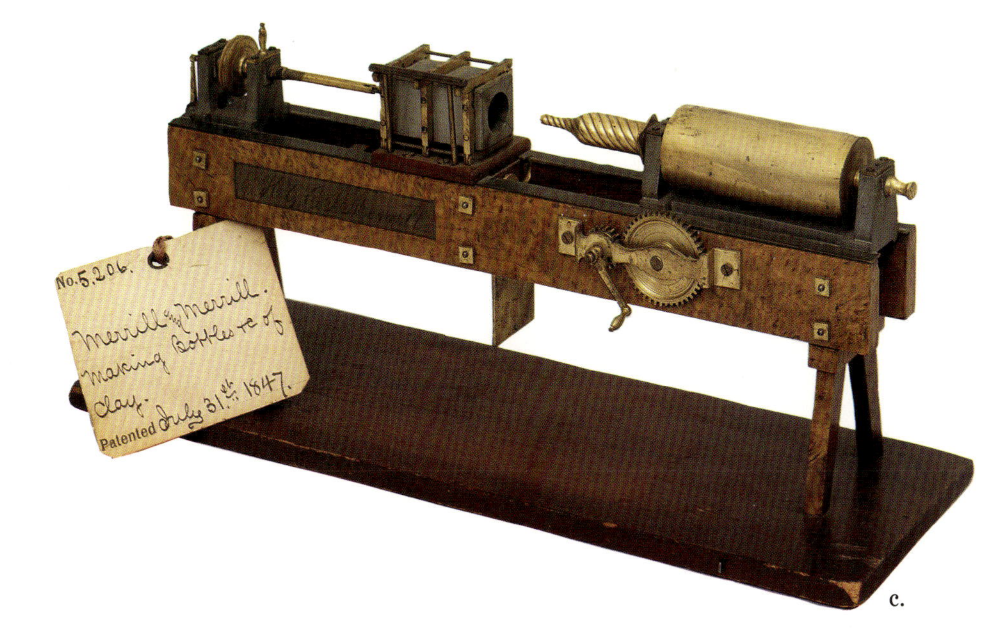

c.

d.

c.
**Improvement in Making Bottles etc.
  of Clay**
Edwin H. Merrill &
  C. J. Merrill
Akron, Ohio
July 31, 1847
Patent No. 5,206
PC, Photograph: Barry Korn

d.
Machine for Sawing Shingles
Henry F. Snyder
Williamsport, Pennsylvania
April 12, 1881
Patent No. 240,048
PC

a.

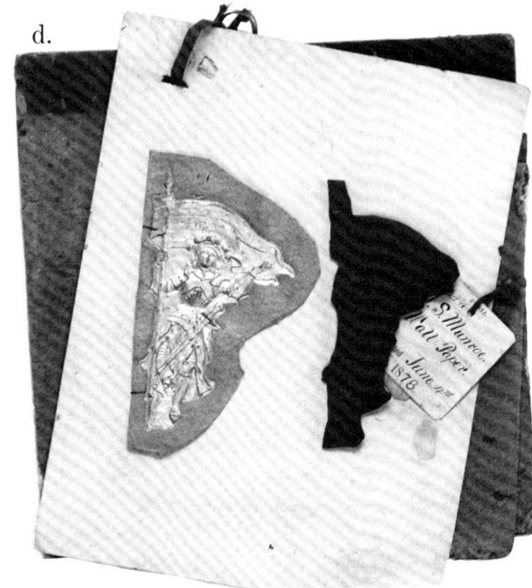

d.

c.

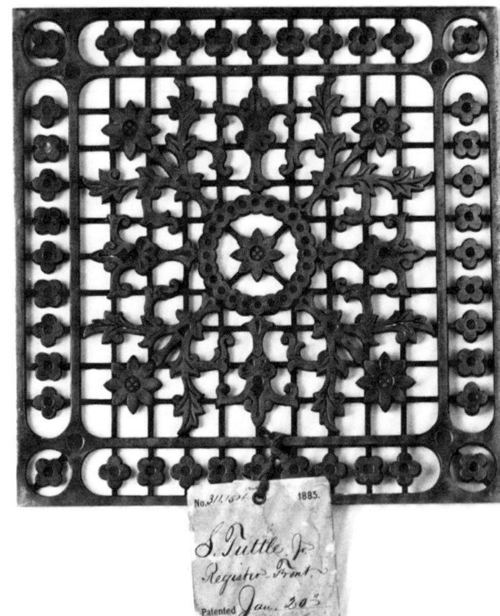

b.

e.

For Architects, Designers, and Builders: Materials, Components, Ornaments, and Details

On these two pages and the following two is a potpourri of patent models—some functional and some decorative (some both), some for the mode of manufacture, others for the manufacture itself. The array of chimney and ventilator caps is especially interesting, as models such as these have rarely been pictured before.

Also noteworthy is the Fugman "Glass Prism-Plate," not so much for its function of refracting "large quantities of light into interiors" as for its date: June 27, 1899, the newest model pictured in this book. Here was a model that was being *added* to the Patent Office collection even as the collection was being consigned to dead storage in an unheated warehouse!

a.
Panel and Wood Trimming
Frederick Mankey
Williamsport, Pennsylvania
August 19, 1884
Patent No. 309,070
PC

b.
Stencil-Plates
Samuel W. Reese
Chicago, Illinois
March 3, 1874
Patent No. 148,087
PC

c.
Register
Silas Tuttle, Jr.
Brooklyn, New York
June 20, 1885
Patent No. 311,154
PC

d.
Improvement in Wall Paper
James S. Munroe
Lexington, Massachusetts
June 19, 1877
Patent No. 204,446
PC

e.
**Improvement in Making Block
   Letters**
Lewis Katen
New York, New York
September 20, 1844
Patent No. 3750
PC

f.
Metal Screen
James Hopkins
New York, New York
November 18, 1862
Patent No. 36,957
PC

g.
**Composition for Imitation of
    Carving**
William B. Gleason
[No number]
PC

h.
Glass Prism Plate
Godfrey Fugman
Cleveland, Ohio
June 27, 1899
Patent No. 627,848
PC

i.
Surface for Printing on Metals
John M. Ronemous
Baltimore, Maryland
October 14, 1879
Patent No. 220,549
PC

j.
Wire Mat
Thomas Midgley
Beaver Falls, Pennsylvania
March 11, 1890
Patent No. 422,960
PC

a.
Improvement in Blind-Hinges
Max Adler
Buffalo, New York
November 3, 1868
Patent No. 83,585
PC

b.
**Improvement in Construction of
  Houses**
Robert G. Lindsay and
  Christian G. Lindsay
Hollidaysburg, Pennsylvania
October 6, 1877
Patent No. 199,076
PC

c.

Improvement in Ventilator Caps
Gerald Kavanaugh
New York, New York
June 30, 1874
Patent No. 152,496
PC

d.

Chimney Cap
William Chappell
Buffalo, New York
October 6, 1868
Patent No. 82,693
PC

e.

Ventilator
David Wells
Lowell, Mass.
November 2, 1852
Patent No. 9,375
PC

f.

Improvement in Chimney Cowls
Andrew F. Barry and
  Ira G. Lane
New York, New York
May 29, 1877
Patent No. 191,297
PC

g.

Improvement in Supply-Ventilators
Jean E. Richard
New York, New York
September 23, 1878
Patent No. 212,056
PC

h.

Improvement in Chimney-Tops
William Richards
London, England
May 2, 1871
Patent No. 114,342
PC

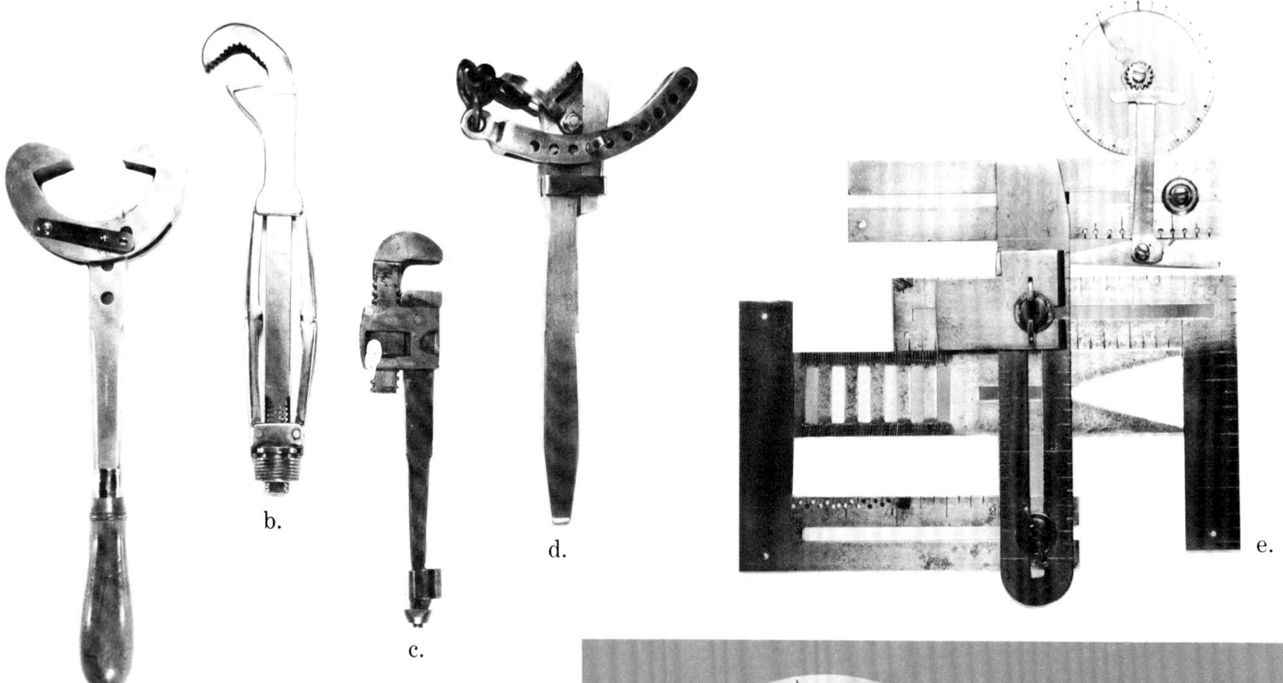

Tools and Implements of the Artisan and Mechanic

a.
Pipe Wrench
O. G. Barrett
Boston Highlands,
   Massachusetts
January 31, 1871
Patent No. 111,422
SI

b.
Pipe Wrench
Seth E. Hurlbut
Chicago, Illinois
December 23, 1879
Patent No. 222,787
SI

c.
Improvement in Wrenches
Daniel C. Stillson
Charlestown, Massachusetts
September 13, 1870
Patent No. 107,304
SI

d.
Improvement in Pipe Wrenches
Royal R. Piper
East Saginaw, Michigan
September 16, 1879
Patent No. 219,758
SI

e.
**Improvement in Bevel and Try-
   Squares**
John Graham
Ludlow, Vermont
November 5, 1867
Patent No. 70,547
SI

f.
Improvement in Levels
Daniel Masten
Binghamton, New York
September 26, 1866
Patent No. 58,271
SI

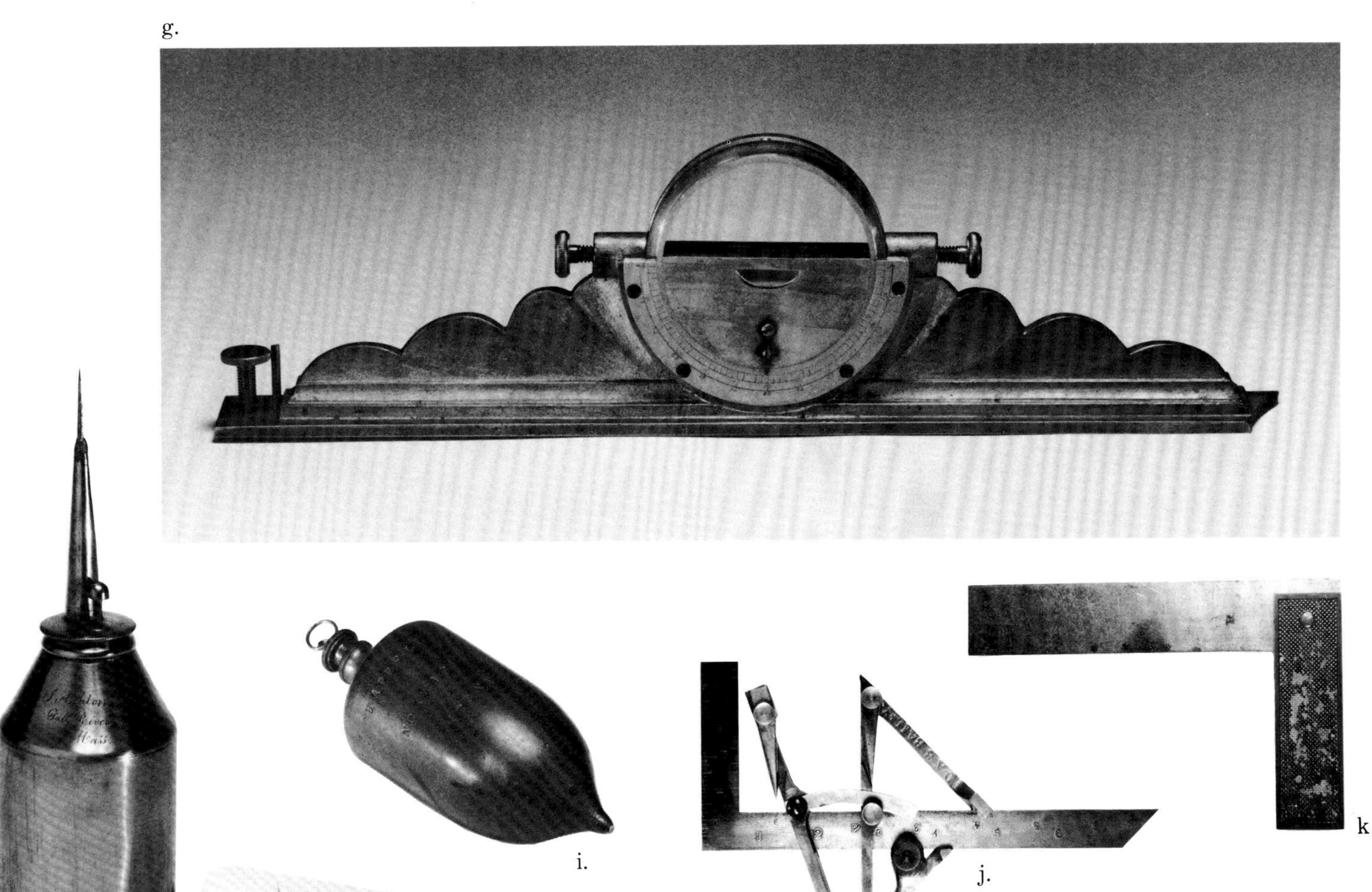

g.

**Improvement in Adjustable
 Spirit Levels**
L. L. Davis
Springfield, Massachusetts
September 17, 1867
Patent No. 68,961
SI

h.

Improvement in Oil Cans
James Ashton
Fall River, Massachusetts
July 3, 1866
Patent No. 55,975
SI

i.

Plumb Bob
Benjamin F. Chappell
Norwich, Connecticut
November 13, 1860
Patent No. 30,612
SI

j.

**Improvement in Combination
 Squares**
D. A. B. Bailey
St. Johnsbury, Vermont
June 11, 1867
Patent No. 65,631
SI

k.

Improvement in Try Squares
Leonard Bailey
New Britain, Connecticut
December 23, 1873
Patent No. 145,715
SI

a.

Mechanical Movements

One eminent historian of technology defines mechanical movements as "machine components—such as gear trains, cams, escapements, and linkages—that produce particular movements within machines." Here, though commencing with a model specifically for a patent on a mechanical movement, is a diverse group of mechanical devices: links, differential hoists, locks and latches, and mechanisms pertinent to clocks and firearms. Although the Patent Office did have a category for "mechanical movements," in some cases it would have classed these inventions otherwise—under the heading of firearms, say, or hoists. This array of models is too rich to address individually. Note, however, the presence of several immediately recognizable names such as James Sargent (Sargent's "cheap, compact, and simple lock" for "doors, drawers, and other articles") and, on pages 110 and 111, Colt and Ericsson, two of the *Men of Progress*. The Colt model shown here was not the one submitted with the "basic" patent of 1836, but rather with a later one of mid-century. The Paterson Arms Manufacturing Co. provided a replacement for the original 1836 model (which was lost in the first Patent Office fire), and it is this replacement that remains in the Smithsonian collection.

As a reminder that not all patent models are inventive "milestones," on pages 108 and 109 are an animal trap and an unlikely combination invention by A. A. Fradenburg and F. C. Goffin, respectively, two gentlemen who seem to be entirely absent from our historical annals.

a.
Improvement in Mechanical Movements
David T. Partlow
Catawba River, South Carolina
September 9, 1879
Patent No. 219,409
SI

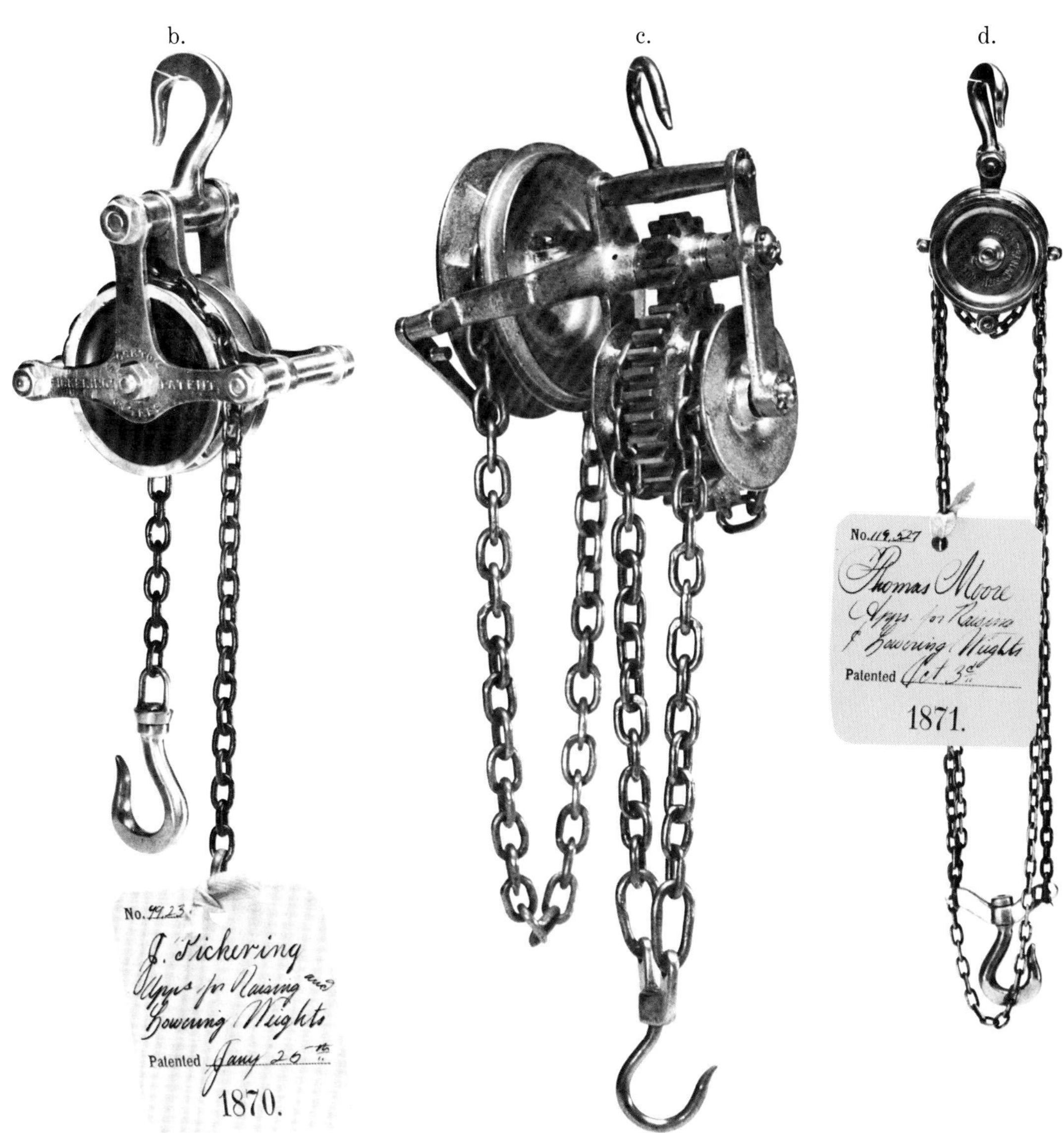

b.
**Improved Apparatus for Raising
 and Lowering Weights**
Jonathan Pickering
Stockton-on-Tees, England
January 25, 1870
Patent No. 99,231
SI

d.
**Improvement in Apparatus for
 Raising and Lowering Weights**
Thomas Moore
Stockton-on-Tees, England
October 3, 1871
Patent No. 119,527
SI

c.
Chain Hoist
[No number]
SI

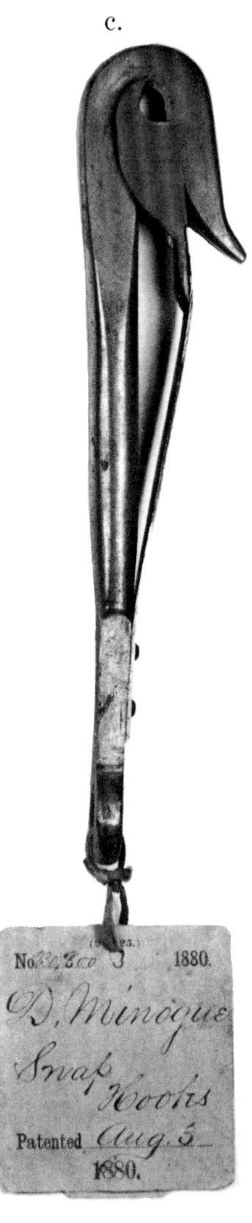

a.
**Improvement in Retaining and
  Releasing Hooks**
Thomas E. Purchase
Reading, Pennsylvania
February 14, 1865
Patent No. 46,385
SI

b.
Improvement in Links
George W. Dyer
Crestline, Ohio
March 12, 1872
Patent No. 124,556
SI

c.
Snap-Hooks
Denis Minogue
Chicago, Illinois
August 3, 1880
Patent No. 230,800
SI

d.
Improved Chain-Link
Samuel Vanstone
Providence, Rhode Island
April 7, 1868
Patent No. 76,562
PC

e.
Chain
Eugene L. Howe
Chicago, Illinois
April 12, 1881
Patent No. 240,132
PC

f.
Improved Link
Alexander Goodhart
Newville, Pennsylvania
May 26, 1868
Patent No. 78,275
PC

g.
Improved Adjustable Link
Andrew J. Dexter
North Fost, Rhode Island
August 2, 1870
Patent No. 105,921
PC

h.
Detachable Link
Samuel Stevens
Alleyton, Michigan
February 27, 1877
Patent No. 187,927
PC

i.
Improvement in Spring-Links
William W. Campbell
Frankfort, Indiana
March 6, 1877
Patent No. 188,100
PC

j.
Openlink
Frederick Schneider
Pagosa Springs, Colorado
October 4, 1881
Patent No. 247,952
PC

a.   b.

d.

c.

a.
Improvement in Compensation Pendulum
Henry B. James
Trenton, New Jersey
October 31, 1871
Patent No. 120,385
SI

b.
**Improvement in
Winding and Setting Watches**
Charles E. Jacot
Chaux-de-Fonds, Switzerland
September 27, 1864
Patent No. 44,493
SI

c.
Clock Escapement
A. D. Crane
Newark, New Jersey
February 16, 1858
Patent No. 19,351
FS

d.
Clock
C. Boardman and Joseph A. Wells
Bristol, Connecticut
January 1, 1847
Patent No. 4,914
SI

e.
Natural Creeping Baby Doll
George Pemberton Clarke
New York, New York
August 29, 1871
Patent No. 118,435
PC, Photograph: Bill Jacobson

a.

b.

c.

a.
**Combined Doorbell and Burglar
  Alarm**
Elliott H. Crane
Burr Oak, Michigan
June 26, 1866
Patent No. 55,823
JF

b.
Improvement in Locks
James Sargent
Rochester, New York
March 11, 1879
Patent No. 203,252
PC

c.
Improvement in Animal Traps
A. A. Fradenburg
Nevada City, California
May 22, 1866
Patent No. 54,885
SI

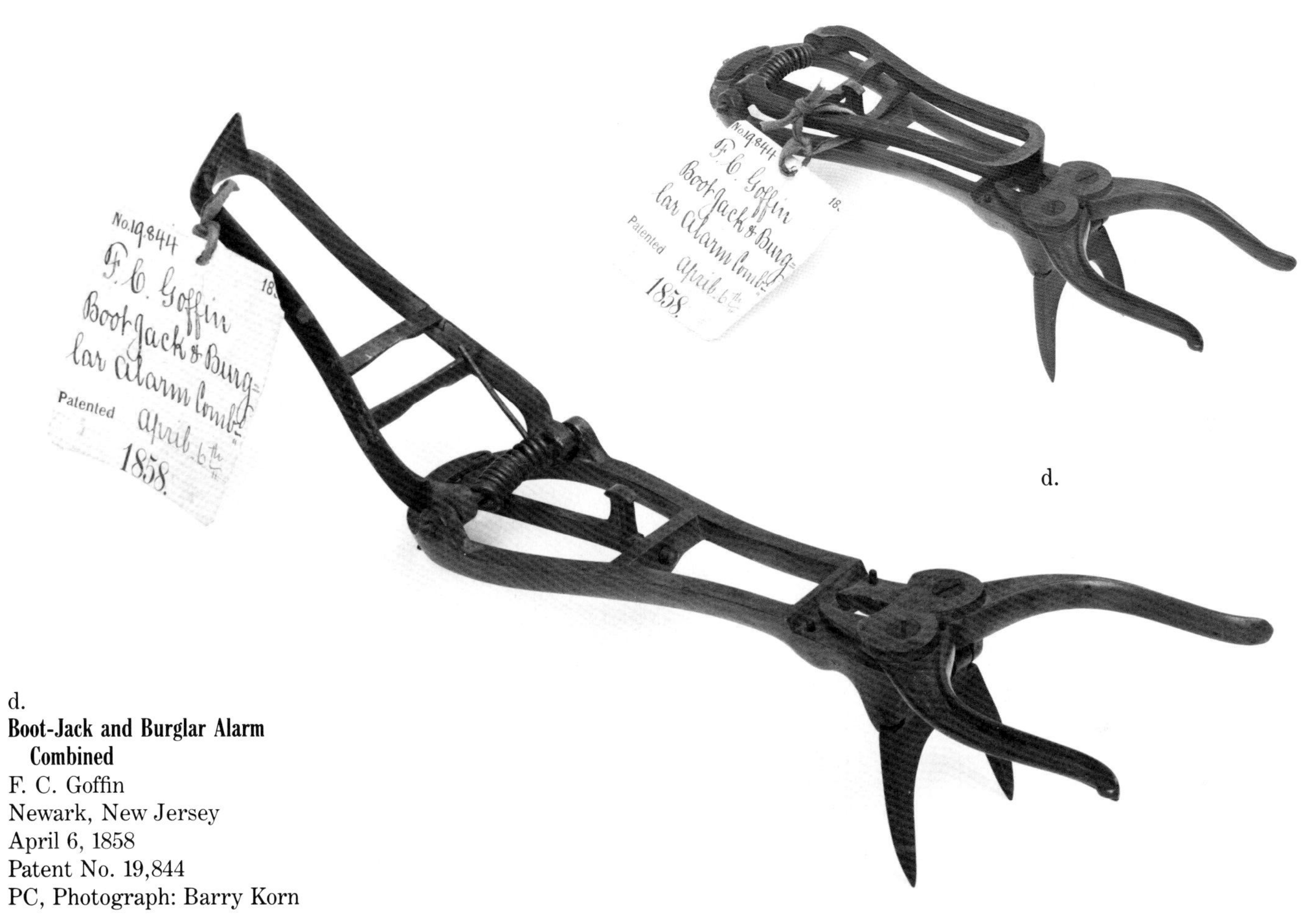

d.
**Boot-Jack and Burglar Alarm
  Combined**
F. C. Goffin
Newark, New Jersey
April 6, 1858
Patent No. 19,844
PC, Photograph: Barry Korn

a.
Gun Carriage
John Ericsson
New York, New York
March 8, 1870
Patent No. 100,514
SI

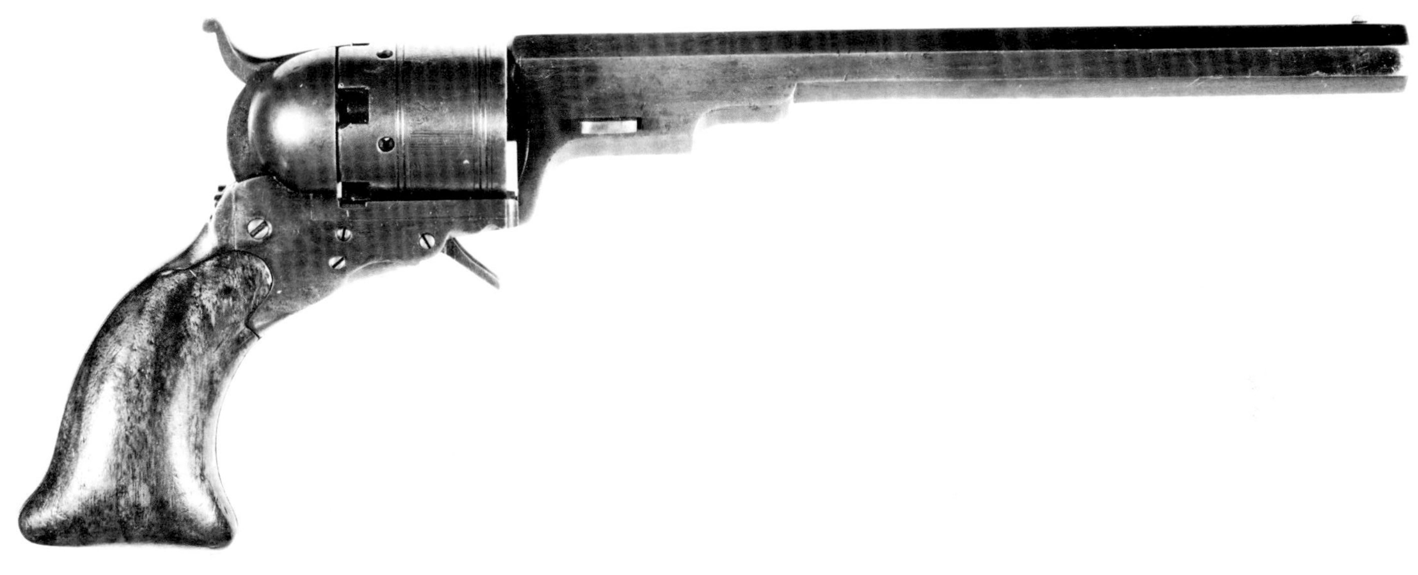

b.
Colt-Paterson Revolver
Samuel Colt
Paterson, New Jersey
February 25, 1836
Unnumbered
SI

a.

b.

c.

Safety

Models on these two pages and the following two pertained to personal safety—Phillips's fire escape, for example; to safety of property—such as Gouley's safe, which could "readily be set afloat . . . in case of danger"; and of course to both together—running lights, headlights, signal lamps, or Abraham Lincoln's "New and Improved Manner of Buoying Vessels Over Shoals." The latter is unquestionably the most famous of all patents never commercialized; and the model, the most famous of *any* model.

Lincoln's model, nicely crafted in wood, stands in contrast to the little tinplate toy submitted by the sons of Elisha G. Otis in conjunction with an application for a patent on a "soft landing" elevator in 1880. The Otis firm was a substantial one, having first made its name in 1853 when the founder demonstrated the safety of his invention at the New York Crystal Palace by riding a platform to the top of a hoistway and then deliberately cutting the rope. Such a model—from which a patent examiner could have learned nothing essential—is evidence that at least one major innovator had come to regard the requirement simply as whimsical.

a.
Signal Headlight for Vessels
William M. & Joseph J. Walton
New York, New York
October 5, 1875
Patent No. 168,434
SI

b.
Police Lantern
James Sheedy
New York, New York
December 8, 1874
Patent No. 157,641
SI

c.
Signal Light for Locomotives
Andrew Dick
Canada
July 23, 1872
Patent No. 129,797
SI

d.
Hoisting Apparatus
Charles and Norton Otis
Yonkers, New York
February 5, 1880
Patent No. 276,673
SI

e.
Fire Escape
William C. Phillips
Norwalk, Connecticut
April 16, 1878
Patent No. 202,460
PC

a.
**New and Improved Manner of
   Buoying Vessels Over Shoals**
Abraham Lincoln
Springfield, Illinois
May 22, 1849
Patent No. 6,489
SI

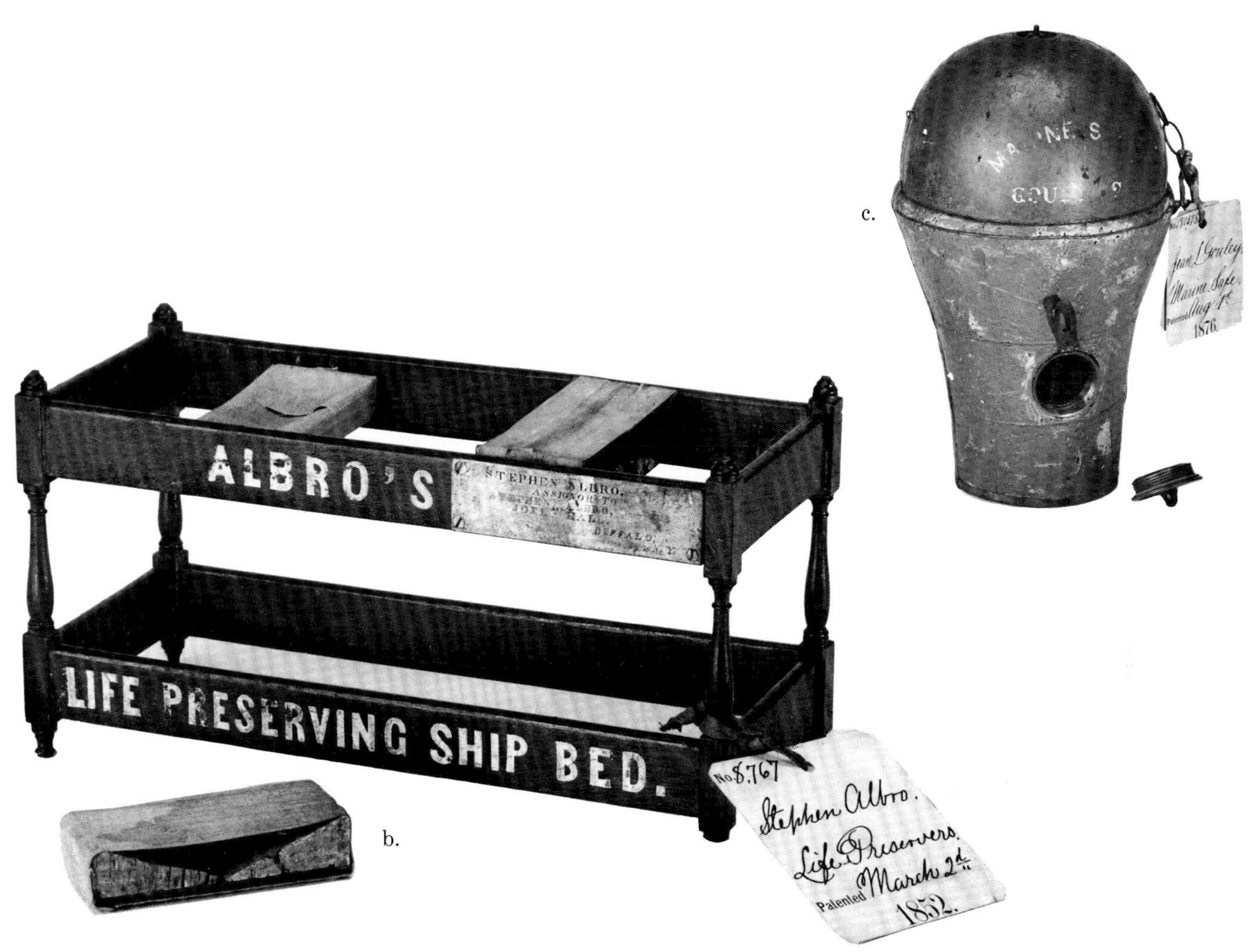

b.
Life Preserving Ship Bed
Stephen Albro
Buffalo, New York
March 2, 1852
Patent No. 8,767
PC

c.
Improvement in Marine Safes
Jean L. Gouley
New Orleans, Louisiana
August 1, 1876
Patent No. 180,575
PC

Safety at Sea

These extraordinary models accompanied patent applications from two men who were contemporaries (Sickels's dates are 1819–1895, Francis's 1801–1893) who specialized in maritime inventions. Both are usually on exhibit in the Smithsonian's Hall of American Maritime Enterprise, and both go far beyond what would have sufficed to indicate the nature of the invention, even to the extent of obscuring it somewhat. Iron ships—the very notion of which had once seemed daft—first appeared in the 1830s, and in the 1840s Francis devised a method of stamping and fabricating corrugated iron lifeboats. The first of these were manufactured by a Brooklyn firm of which he was a co-owner, the Novelty Iron Works, but business was so good that in 1852 Francis established a separate firm, the Metallic Life-Boat Company. The model here accompanied the application for the last of Francis's several patents.

As for the steam steering gear, the idea was to improve the maneuverability of ocean-going vessels, which had been getting both faster and larger, and hence increasingly difficult to manhandle. Sickels produced a prototype of his invention, in effect the first "power steering," which apparently worked well enough in trials aboard several different vessels, although it was never a commercial success.

a.
Boats, Metallic
Joseph Francis
New York, New York
March 23, 1858
Patent No. 19,693
SI

b.
**Operating and Controlling the
  Rudders of Steam Vessels**
Frederick E. Sickels
New York, New York
May 10, 1853
Patent No. 9,713
SI

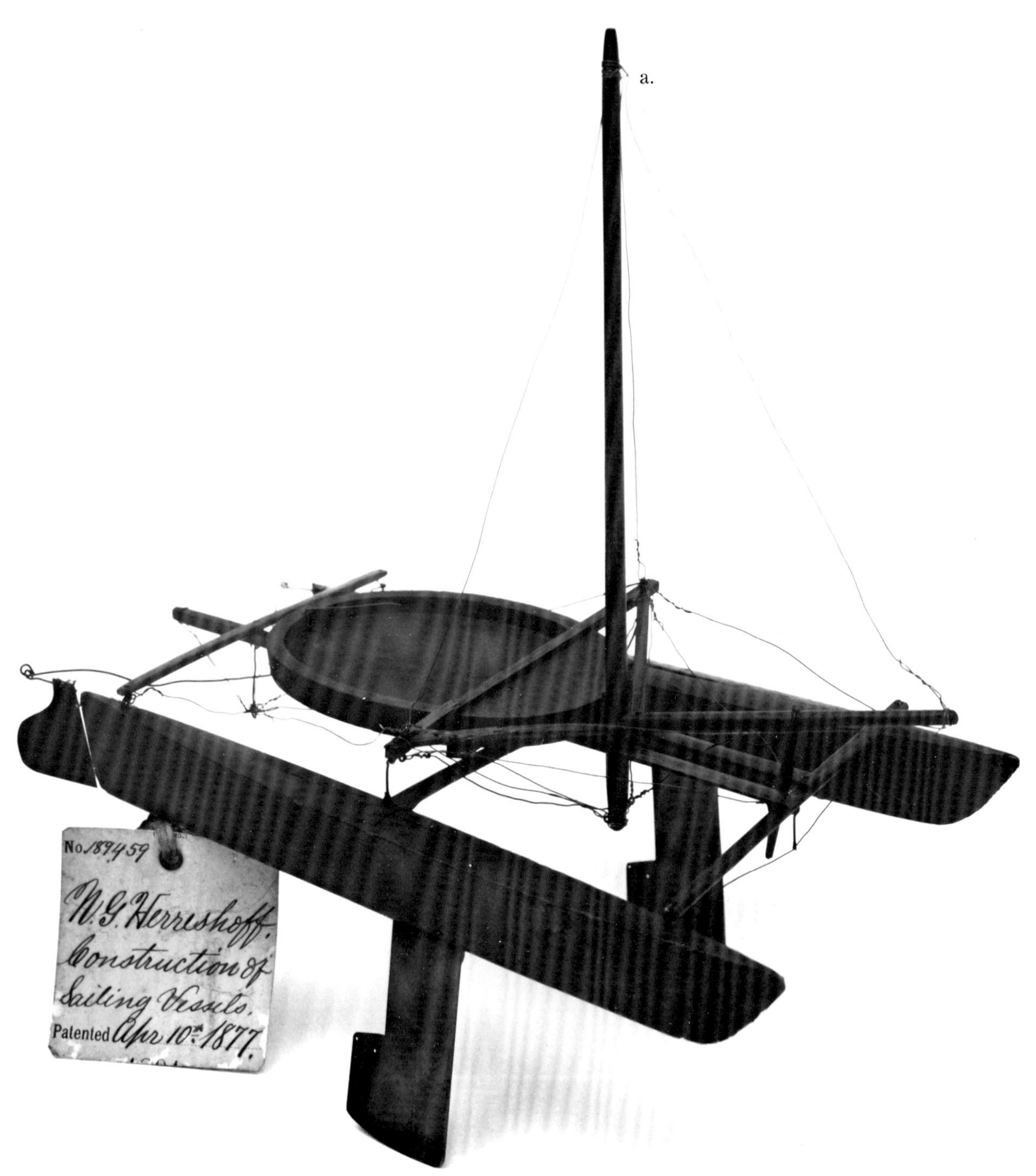

a.
Construction of Sailing Vessel
Nathaniel G. Herreshoff
Providence, Rhode Island
April 10, 1877
Patent No. 189,459
PC, Photograph: PC

b.

c.

d.

Maritime Engineering

On these pages are the names of men whose inventions evidently made little impression even in their own day. Hidden's was a simple improvement to a commonplace item of hardware; Williams's was theoretically intriguing but impractical. A more familiar family name is one that makes national news at least as frequently as each *America*'s Cup competition. Nathaniel Greene Herreshoff patented his unique and very fast catamaran the year before he and his two brothers founded the Herreshoff Manufacturing Co.; he was in charge of drafting, engineering, and experimentation. This firm was the first to build yachts over molds (keel up), with double skins and iron floors and knees, and it became the creator of first-rate racing boats, with a .900 record of victories against the fastest from England.

b.

Improvement in Feathering Paddle-Wheels
Fletcher Felter
Perth Amboy, New Jersey
November 20, 1854
Patent No. 11,992
SI

c.

Side-Lights for Ships
Enoch Hidden
New York, New York
June 21, 1853
Patent No. 9,811
SI

d.

Tug for Towing Boats
Walter Everson
New York, New York
January 3, 1871
Patent No. 110,754
SI

a.

Civil Engineering

The names of few nineteenth-century American bridge designers remain common currency—certainly not Jarvis or Sullivan, Kessler or Foster —but there are exceptions: James B. Eads and, especially, John A. Roebling. An immigrant from Thuringia, Roebling had settled on a Pennsylvania farm in 1831. His natural calling was as an engineer, and he elected to specialize in the design and manufacture of wire rope (cables), in demand for haulage on canal inclines. The model here dates from the period of Roebling's entry into the wire rope business, two years before he built his first suspension bridge, and nine years before the "quantum leap" involved in building the Niagara Bridge, which was under construction from 1851 to 1855.

In 1848, at the behest of Peter Cooper, whose myriad enterprises included the manufacture of wire, Roebling relocated his works in Trenton, New Jersey, and that firm remained in business until the late 1950s. Roebling's immortality, of course, rests essentially in his design of the "Great East River Bridge"— which has become the most storied of all monuments to American technological enterprise—but it rests too in a scattering of other artifacts, including this little wood and brass model, which Roebling probably built with his own hands.

a.
Machine for Making Wire Rope
John A. Roebling
Sachsenburg, Pennsylvania
July 16, 1842
Patent No. 2,720
SI

b.
Bridges
Philip Jarvis
Mount Ayr, Iowa
March 2, 1879
Patent No. 212,941
SI

c.
Wooden Truss Bridge
Mark J. Sullivan, J. Kessler,
    and J. R. Foster
Dubuque, Iowa
February 10, 1880
Patent No. 224,491
SI

Locomotive Engineering

On July 4, 1828, the aged Charles Carroll—the last surviving signer of the Declaration of Independence—laid the cornerstone for the nation's first railroad, the Baltimore & Ohio. Steam power came to the B&O two years later when Peter Cooper, who had diversified from glue into iron, put his *Tom Thumb* into service in Baltimore. *Tom Thumb* was less than an unqualified success, however, and not until a year later did the *York* steam locomotive prove fully serviceable on the B&O. By then the chief engineer of another railroad in the South, a New Yorker named Horatio Allen, was already operating his *Best Friend of Charleston* regularly.

Before the emergence of the railroads as America's first "big business," inventors—and sometimes investors—showed a marked penchant for "new-fangled" designs. Indeed, Patent No. 1 was issued to John Ruggles, the Senator who drafted the Patent Act of 1836, for a locomotive of "multiplied tractive power." By mid-century, experimentation was rife,

there being no better example than the express locomotive with a single pair of ten-foot drivers patented by one of Ruggles's constituents from Maine, Richard Emerson. Emerson's patent included several claims, but its originality was borderline at best, and its practicality nil. All this suggests that applications were being approved without a very thorough examination; in the 1850s, by contrast, the examiner specializing in railroad apparatus was so strict about searching out "prior art" that few new patents were issued even as the railroad business achieved an incredible dynamism, with both mileage and investment tripling—the latter to $1.15 billion by the time of the Civil War.

While Emerson's high-speed ideal never materialized, his tin, iron, and sheet-lead patent model with its whimsical smokestack and geometric spokes remains a delightful relic. Another approach to the

design of a model is exemplified in Cathcart's rack locomotive, a device which at full scale would have been largely of metal but which is here done in wood. Cathcart's invention differed, too, in that it was actually produced for service on the Madison & Indianapolis Railroad, whose rails ascended a 6 percent incline from the banks of the Ohio. As Cathcart was solving a local problem in southern Indiana, in Cincinnati George E. Sellers was attempting to promote another sort of grade-climbing locomotive, this one for operating on lines to be built with only a minimum of expensive grading. When the natural lay of the land was too steep for conventional locomotives, steam-driven "gripper" wheels would press against a third rail between the other two, thus providing a sufficient assist.

Ultimately, Sellers's idea came to no more than Emerson's had—the railroads he had in mind simply were not built—but, again, we are left with a remarkable model,

which, save for the key gripper system inside the drivers, is otherwise rendered as an abstraction. In later years, models for patents on discrete locomotive components were usually rendered in terms of that component alone (see the spark arrestors on the next page); though with less complex machines, miniatures were conventional—the handcars, for example. The station indicator is no doubt full-scale or close to it.

c.

a.

**Improvement in the Boilers and
Gearing of Locomotive Engines
for Working Heavy Grades**

George E. Sellers
Cincinnati, Ohio
July 9, 1850
Patent No. 7,498
SI

b.

**Locomotive With Driving Axle
Above the Boiler**

Richard H. Emerson
Portland, Maine
May 1849
Patent No. 6,401
SI

c.

**Improvement in Locomotives for
Ascending Inclined Planes**

Andrew Cathcart
Madison, Indiana
October 23, 1849
Patent No. 6,818
SI

123

b.

c.

a.

d.

a.
Railroad Switch Signal
Thomas Daly
Erie, Pennsylvania
May 18, 1869
Patent No. 90,241
DF

b.
Improvement in Railway Hand-Cars
George S. Sheffield
Three Rivers, Michigan
March 11, 1879
Patent No. 213,254
SI

c.
Hand-Car
Thomas G. Glover, Jr.
Bedford, Indiana
April 6, 1880
Patent No. 226,304
SI

d.
**Improvement in Spark Arresters
    for Locomotives**
Charles S. and Edward Osborn
Newton, New Jersey
March 9, 1875
Patent No. 160,614
SI

e.

e.
Station Indicator
John Mulligan
New York, New York
February 24, 1874
Patent No. 147,859
PC

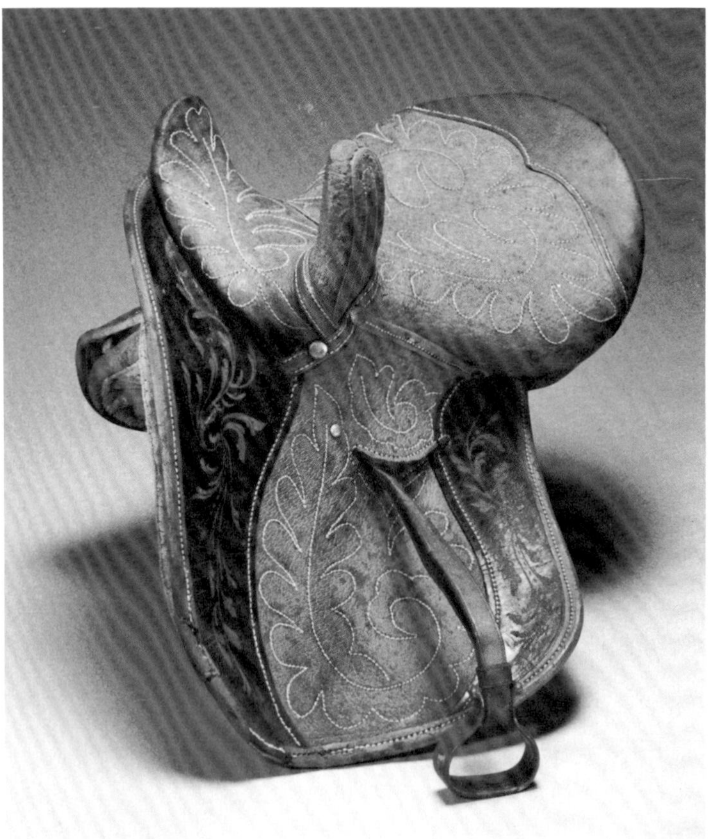

Overland Transportation

In a country with only a rudimentary system of public roads, and where the alternative of canals was exceedingly costly, adoption of the technology of iron wheels on iron rail marked a tremendous advancement. Inventors continued, however, to ponder ways of improving more traditional modes of overland transportation via horseback or horse-drawn vehicles, at least in small ways.

Bicycles and tricycles, initially popularized in England, began to attract attention in the U.S. after the Centennial. In 1878 Albert A. Pope contracted to manufacture an American version of the high-wheeler, and the Pope "Columbia" marks the takeoff of the bicycle era. The "safety" bicycle with differential sprockets appeared in the late 1880s, precipitating the bicycle "craze" of the following decade.

Meantime, the internal-combustion automobile had begun to make its presence felt, and there had likewise been efforts —since the eighteenth century, in fact—to develop road carriages powered by steam. In terms of design, J. E. Praul's was a less likely candidate for success than many contrivances long antedating his—the old idea of propulsive "feet" was as hopeless as ever—but the survival of his model is certainly a happy circumstance.

a.
Improved Riding-Saddle
George Horter
New Orleans, Louisiana
July 5, 1870
Patent No. 105,080
SI

b.
Improved Hitching-Strap
A. J. Ross
Rochester, New York
November 5, 1857
Patent No. 70,473
PC

c.
Traction Engine
J. E. Praul
Washington, D.C.
September 26, 1879
Patent No. 221,354
SI

127

a.
Wheel and Axle
John A. David [for] Soule and
    Manuel
Hyde Park, Massachusetts
May 10, 1881
Patent No. 241,244
JF

b.
Wheel Fender for Vehicles
M. Wheeler
Philadelphia, Pennsylvania
June 13, 1882
Patent No. 259,619
JF

c.
Tricycle
Francis Fowler
New Haven, Connecticut
February 3, 1880
Patent No. 224,165
SI

d.
Tricycle
Otto Unzicker
Chicago, Illinois
June 4, 1878
Patent No. 204,636
SI

e.
Improved Bicycle
William Klahr
Myerstown, Pennsylvania
March 5, 1883
Patent No. 285,821
SI

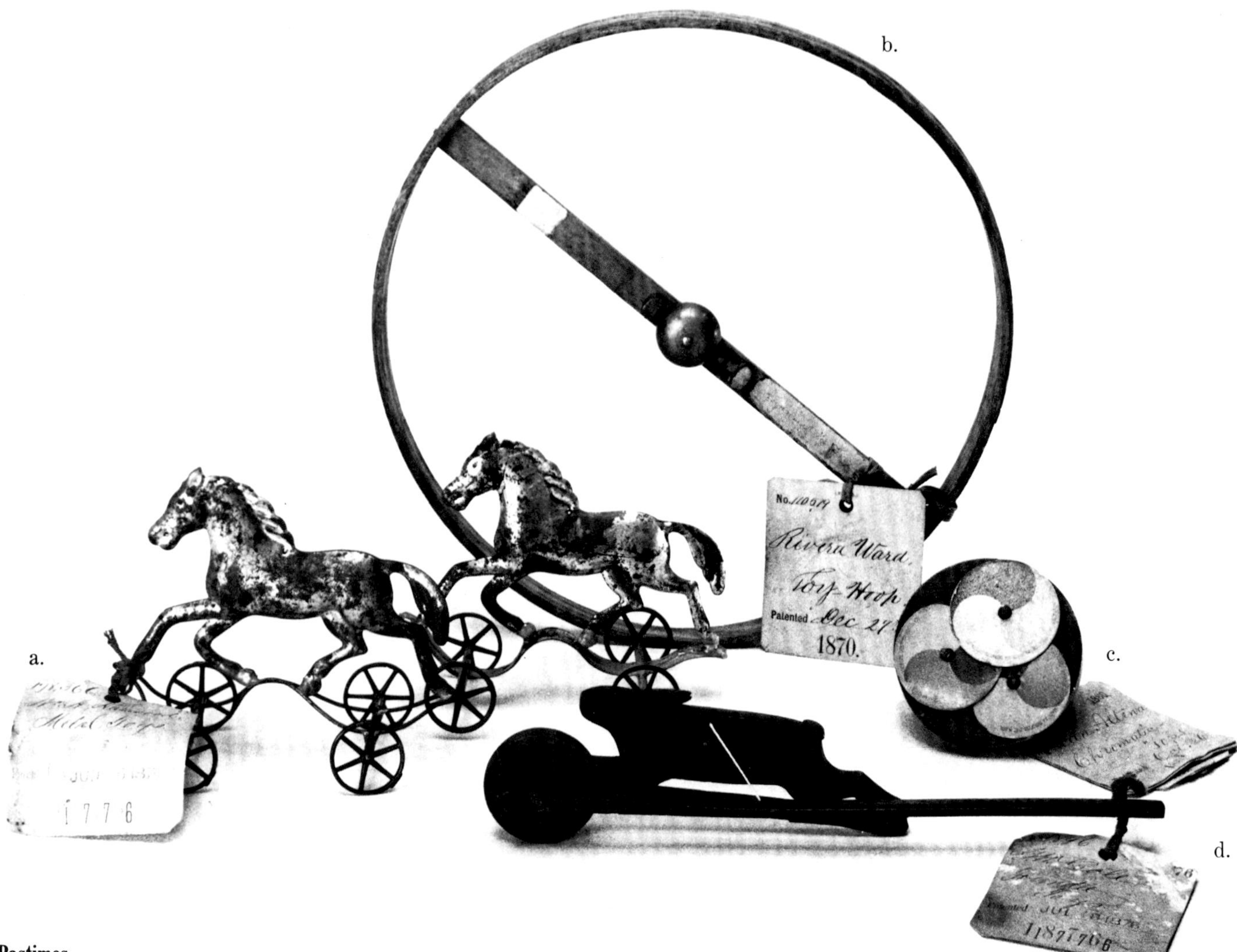

Pastimes

On these concluding pages are models pertaining to patents that had to do with entertainment, games, recreation, and leisure-time activities. These include toys such as hoops and tops, a wind-up acrobat, and tin horses on wheels. On the next two pages are some paraphernalia of "socializing," including lawn furniture, a swing, ice skates, and a "parlor" skate whose design facilitated maneuverability and "many graceful revolutions" while "reducing the likelihood of collisions." These are followed by some items of novelty furniture, accoutrements of musical and theatrical performances, and lastly an endearing accessory to a test of marksmanship.

Most are models in the sense of being examples rather than miniatures, and even the furniture could actually be full-scale toys. There is only one patent of any special technological interest, that of Christian Frederick Theodore Steinway. Unlike his father and brother, Christian was only in the U.S. from 1865 to 1870, a brief but eventful interval. During that time, Steinway Hall formally opened, and intensive experimentation in the application of acoustical science to piano design resulted in this patent on a method of obtaining "harmonic intermediate tones of vibration, which add fulness and roundness to the increased prolongation of sound realized by constructing the bridge with free-suspended portions."

The latter few words could well have been part of a specification for a patent in the realm of civil engineering, and definitely this is a patent for something of considerable sophistication. While there is a full measure of models for such patents in this book, the fundamental aim has been to suggest the diversity of this species of artifact. Relatively few pertain to inventions that had anything to do with "having altered the course of civilization" (to borrow the phrase from the *Men of Progress* one last time). But all together, the survivors may serve an important function today, as a way of perceiving a style of creative ingenuity indispensable to an understanding of nineteenth-century America.

e.

a.
Improvement in Metal Toys
William A. Harwood
Brooklyn, New York
June 6, 1876
Patent No. 178,366
DF

b.
Improvement in Toy-Hoops
Rivera Ward
Newark, New Jersey
December 27, 1870
Patent No. 110,519
PC

c.
Chromatic Top
Henry Van Altena
Milwaukee, Wisconsin
October 26, 1880
Patent No. 233,820
PC

d.
Improvement in Trundle Toys
Gideon W. Cole
Canton, Illinois
July 18, 1876
Patent No. 179,896
PC

e.
Automatic Toy
Henry L. Brower
New York, New York
July 15, 1873
Patent No. 140,883
PC, Photograph: Barry Korn

131

a.
Improvement in Lawn-Seats
William C. Medcalf
Rochester, New York
September 16, 1873
Patent No. 142,927
SI

b.
Improvement in Swings
George Sill
Wilkins, Pennsylvania
February 2, 1869
Patent No. 86,598
PC

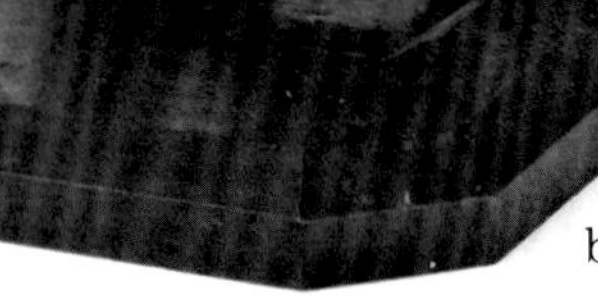

b.

c.

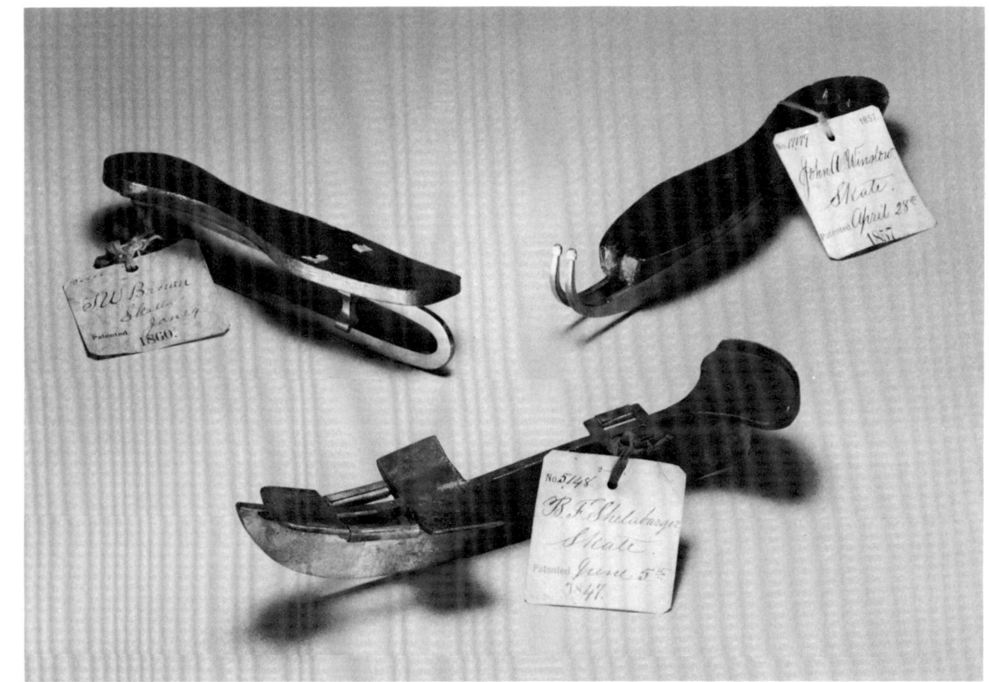

c.

Improvement in Parlor Skates
Hiram L. True
McConnellsville, Ohio
April 3, 1877
Patent No. 189,285
PC, Photograph: Barry Korn

d.
Skate
Thomas W. Brown
Boston, Massachusetts
January 24, 1860
Patent No. 26,882
SI

e.
Skate
John A. Winslow
Roxbury, Massachusetts
April 28, 1857
Patent No. 17,179
SI

f.
Skate
B. F. Shelabarger
Mifflintown, Pennsylvania
June 5, 1847
Patent No. 5,138
SI

a.
Musical Rocking Chair
Clayton Denn
Philadelphia, Pennsylvania
August 30, 1870
Patent No. 106,790
PC, Photograph: Barry Korn

b.
Rotating Blast-Producing Chair
L. R. Breisach
New York, New York
February 16, 1858
Patent No. 19,343
PC, Photograph: Barry Korn

a.
Piano-Tuning Pins
John Lauterschlager
Ashland, Kentucky
February 12, 1878
Patent No. 200,207
PC

b.
**Improvement in Sound-Boards for
  Piano-Fortes**
Christian Frederick Theodore
  Steinway
New York, New York
April 6, 1869
Patent No. 88,749
PC

c.

c.
Theatre-Stages
Marshall H. Mallory
New York, New York
June 8, 1880
Patent No. 228,468
PC

a.
Improvement in Game Apparatus
John Brown
Providence, Rhode Island
November 13, 1877
Patent No. 197,091
JF

b.
Improvement in Pigeon Starters
Henry A. Rosenthal
Brooklyn, New York
February 16, 1875
Patent No. 159,846
PC, Photograph: PC

Suggested Readings

Arnold, Martin. "100,000 Antique Patent Models Stacked in Barns." *New York Times*, May 31, 1971.

Brown, D. W. *Brief History of Patent Legislation in the United States*. New York, 1889.

Bugbee, Bruce W. *Genesis of American Patent and Copyright Law*. Washington, D.C., 1967.

Campbell, Levin H. *The Patent System of the United States So Far As It Relates to the Granting of Patents: A History*. Washington, D.C., 1891.

Caplan, Ralph. *By Design*. New York, 1982.

Dood, Kendall. "Patent Models and the Patent Law, 1790–1880." *Journal of the Patent Office Society*, April 1983 (Part 1), pp. 187–216; May 1983 (Part II), pp. 234–47.

Federico, P. J., editor. "Outline History of the United States Patent Office." *Journal of the Patent Office Society*, Vol. 18, 1936, pp. 1–136.

Ferguson, Eugene S. "The Mind's Eye: Nonverbal Thought in Technology." *Science*, August 26, 1977, pp. 827–36.

Ferguson, Eugene S. and Christopher Baer. "Little Machines: Patent Models in the Nineteenth Century." The Hagley Museum, Greenville, Delaware, 1979.

Gilfillan, S. C. *The Sociology of Invention: An Essay in the Social Causes of Technic Invention and Some of Its Social Results*. Chicago, 1935.

Greenberg, Jan, Jessica Hageman, and William Schinsky. "American Patent Models 1836/1880." The Art Gallery, California State University at Fullerton, 1977.

Hindle, Brooke. *Emulation and Invention*. New York, 1981.

Hogan, Donald W. "Unwanted Treasures of the Patent Office." *American Heritage*, February 1957, pp. 16–19.

Horwitz, M. J. *The Transformation of American Law, 1780–1860*. Cambridge, Mass., 1977.

Imlow, E. B. *The Patent Grant*. Baltimore, 1950.

Mayr, Otto, and Robert C. Post, editors. *Yankee Enterprise: The Rise of the American System of Manufactures*. Washington, D.C., 1981.

Mitchell, Charles Eliot. "Birth and Growth of the American Patent System," in *Celebration of the Beginning of the Second Century of the American Patent System*. Washington, D.C., 1892, pp. 43–55.

Nelson, George. *George Nelson on Design*. New York, 1979.

Nelson, George. *How to See: A Guide to Reading our Man Made Environment*. Boston, 1979.

Nelson, George. *Problems of Design*. New York, 1965.

Odell, Rice. "Patent Models Recall Creativity of Early America." *Smithsonian*, April 1973, pp. 59–65.

Post, Robert C. "The American Genius," in *The Smithsonian Book of Invention*, ed. Post. New York, 2nd ed., 1978, pp. 22–31.

———. "From Pillar to Post: The Plight of the Patent Models." *IA: The Journal of the Society for Industrial Archeology*, 1978, pp. 58–60.

———. "'Liberalizers' Versus 'Scientific Men' in the Antebellum Patent Office." *Technology and Culture*, January 1976, pp. 24–54.

———. *Physics, Patents, and Politics: A Biography of Charles Grafton Page*. New York, 1976.

Pye, David. *The Nature of Design*. New York, 1964.

Ray, William and Marlys. *The Art of Invention: Patent Models and Their Makers*. Princeton, N.J., 1974.

Reingold, Nathan. "U.S. Patent Office Records as Sources for the History of Invention and Technological Property." *Technology and Culture*, Spring 1960, pp. 156–167.

Robinson, W. C. *The Law of Patents for Useful Inventions*. Boston, 1890.

Rossman, Joseph. *Industrial Creativity: The Psychology of the Inventor*. New Hyde Park, N.Y., 1964.

Schmookler, Jacob. *Invention and Economic Growth*. Cambridge, Mass., 1966.

Scientific American. *The Scientific American Reference Book: A Compendium of Useful Information for Inventors and Mechanics, Comprising the Patent Laws of the United States . . . & c.* New York, 1876.

Sherwood, Morgan. "A Patent Madness," in *The Smithsonian Book of Invention*, ed. Robert C. Post. New York, 2nd ed., 1978, pp. 156–59.

———. "The Origins and Development of the American Patent System." *American Scientist*, September–October 1983, pp. 500–506.

Trillin, Calvin, "U.S. Journal: Garrison, N.Y. Jackpot." *The New Yorker*, January 7, 1974, p. 52.

United States Patent Office. *An Account of the Destruction by Fire of North and West Halls of the Model Room in the United States Patent Office Building, On 24th of September, 1877, Together With a History of the Patent Office from 1790 to 1877*. Washington, D.C., 1877.

Welsh, Peter C. *United States Patents, 1790 to 1870: New Uses for Old Ideas (Contributions From the Museum of History and Technology, United States National Museum Bulletin 241)*. Washington, D.C., 1966.

Wyman, William I. "The Patent Act of 1836." *Journal of the Patent Office Society*, Vol. 1, 1919, pp. 203–10.

Chronology

1787
Patent protection clause included in the U.S. Constitution

1790
First patent act passed by Congress; patent commission established; patentees required to submit model, drawing(s), and description with each application

First patent granted by U.S. government (to Samuel Hopkins for a process of making pot ash and pearl ashes)

1793
Congress passes new patent act: patent commission abolished, and Secretary of State given power to grant patents; requirement for model rescinded

1800
Washington, D.C. established as capital of the U.S.

Patent Office temporarily set up within the Treasury Office

1802
Patent Office again attached to the State Department

1810
Patent Office set up in Blodgett's Hotel, next to the City Post Office

1823
Inventory taken by Patent Office; 1,819 patent models accounted for

1835
Henry L. Ellsworth is appointed as Patent Superintendent by President Jackson

A congressional committee chaired by Senator John Ruggles begins investigating the deteriorating patent system

1836
Ruggles's committee issues report; Congress passes two new patent acts on July 4, establishing a rigorous examination process and legally requiring models with all patent applications; new patent office building proposed

Patent No. 1 issued on July 13 (to Sen. Ruggles for traction wheels for locomotive steam-engine). From 1790 to 1836, there were 9,957 unnumbered patents issued)

Fire breaks out in Blodgett's Hotel in December, destroying most of the Patent Office and almost all models

1837
Congress appropriates money to replace patent office records and some of the models destroyed in the fire

1840
Patent Office Building on 8th Street, N.W., completed and occupied

1842
Patent system begins to include designs under its protection; first design patent issued (to George Bruce for a typeface)

1844
Congress appropriates money for enlarging Patent Office Building

1852
East wing of Patent Office Building constructed

1857
West wing of Patent Office completed

Number of patents granted since 1836 reaches 10,000 (20,000 since 1790)

1860
North wing added to Patent Office Building, completing the quadrangle of the building's design; Patent Office occupies it after the Civil War, in 1867

1861
Congress authorizes Patent Commission to dispose of models that accompanied rejected patent applications and models for design patents

1870
Congress passes act rescinding requirement for models (but permitting Commissioner to request models in certain cases)

1871
Number of patent models reaches 200,000

1877
Fire severely damages west and north wings of Patent Office Building; at least 76,000 models destroyed

1880
Number of patent models reaches 250,000

Patent Office eliminates requirement for models once and for all (but models continue to be submitted by patentees)

1908
Patent Office first attempts to place the models elsewhere; Smithsonian Institution takes over 1000.

1911
Patent No. 1,000,000 issued

1925
Several thousand more models go to the Smithsonian Institution, the Henry Ford Museum and other institutions; more than 50,000 are sold at public auction

1926
Sir Henry Wellcome, expatriate pharmaceutical magnate, purchases the remaining models (100,000 +) for $6,540, with the intention of establishing a museum of invention in Tuckahoe, N.Y.

1936
Wellcome's collection sold for $50,000 to a New York theatrical producer

1942
The collection is sold again (at bankruptcy auction), to O. Rundle Gilbert, an upstate New York auctioneer, who begins selling them in auctions and through department stores

1945
Fire destroys 20,000 of Gilbert's models in storage

1962
Patent No. 3,000,000 issued

1975
Name of Patent Office changed to the Patent and Trademark Office by Congress

1979
The models remaining in Gilbert's collection (perhaps 30,000) are bought by Cliff Petersen, a California aerospace engineer